Fractional Factorial Plans

WILEY SERIES IN PROBABILITY AND STATISTICS
PROBABILITY AND STATISTICS SECTION

Established by WALTER A. SHEWHART and SAMUEL S. WILKS

A complete list of the titles in this series appears at the end of this volume.

Fractional Factorial Plans

ALOKE DEY
Indian Statistical Institute
New Delhi, India

RAHUL MUKERJEE
Indian Institute of Management
Calcutta, India

A Wiley-Interscience Publication
JOHN WILEY & SONS, INC.
New York • Chichester • Weinheim • Brisbane • Singapore • Toronto

This book is printed on acid-free paper. ♾

Published simultaneously in Canada.

Library of Congress Cataloging-in-Publication Data:

Dey, Aloke.
Fractional factorial plans / Aloke Dey, Rahul Mukerjee.
p. cm. — (Wiley series in probability and statistics. Probability and statistics section)
"A Wiley-Interscience publication."
Includes bibliographical references and index.
ISBN 0-471-29414-4 (cloth)
1. Factorial experiment designs. I. Mukerjee, Rahul. II. Title. III. Series: Wiley series in probability and statistics. Probability and statistics.
QA279.D467 1999
519.5—dc21 98-45208

Printed in the United States of America

10 9 8 7 6 5 4 3 2 1

To my wife and daughters, AD

To my parents and wife, RM

Contents

Preface

Fractional factorial plans are of immense practical utility in many fields of investigation and research in this area is progressing at a vigorous pace. The literature is already voluminous and continues to grow. This volume aims at presenting a cohesive, up-to-date and mathematical treatment of the theory of fractional factorials.

The present book makes an extensive application of a Kronecker calculus for factorial arrangements. The use of this calculus, among other things, helps in developing a unified theory and, in some cases, notably in Chapters 2 and 5, simpler proofs. Much of the material is available for the first time in the form of a book.

The book is organized into eight chapters followed by an appendix; a brief description of the contents appears in Chapter 1, which is of introductory nature. While Chapters 2, 6, 7, and 8 deal with optimal fractions from various standpoints, the related combinatorial theory of orthogonal arrays has been reviewed in Chapters 3, 4, and 5.

We assume a background in basic matrix algebra and linear statistical models. Chapters 1 and 4, respectively, of *Linear Statistical Inference and its Applications*, Second Edition, by C. R. Rao provide adequate preparation for these. Some familiarity with the general area of experimental design at an advanced undergraduate level will also be of advantage while reading the book. Several references to this end have been provided in Chapter 1.

We have attempted to cover the developments which in our perception are the most important ones. However, the book is not intended to be encyclopaedic. For example, our coverage of orthogonal arrays has been only to the extent to which these are relevant to fractional factorials.

During the course of writing this book, we received encouragement and help from several of our colleagues. We sincerely thank all of them. We are particularly indebted to C. F. Jeff Wu of the University of Michigan for "planting the seed" in our minds to take up this assignment and for his words of constant encouragement. We would also like to express our deep appreciation to Steve Quigley and the editorial staff of John Wiley & Sons for their prompt and excellent handling of the project. Preliminary drafts of the book were read and commented upon by several reviewers. To all of them we express our sincere thanks. We acknowledge the support received for this project from the Centre for Management and Development Studies, Indian Institute of Management, Calcutta. We are also thankful to the Indian Statistical Institute for providing us with a conducive environment to carry out this work.

ALOKE DEY
RAHUL MUKERJEE

New Delhi
Calcutta
January 1999

CHAPTER 1

Introduction

1.1 INTRODUCTORY REMARKS

Factorial experiments have wide applications in many diverse areas of human investigation. Typically in such an experiment there is an output variable which is dependent on several controllable or input variables. These input variables are called *factors*. For each factor there are two or more possible settings known as *levels*. Any combination of the levels of all the factors under consideration is called a *treatment combination*. Factorial experiments aim at exploring the effects of the individual factors and perhaps their interrelationship as well.

A factorial experiment where each treatment combination is applied to at least one experimental unit is called a *complete* factorial. However, quite often in practice, the total number of treatment combinations is too large to allow the use of a complete factorial. Since each factor involves at least two levels, this can happen even with only a moderate number of factors. For example, consider an agronomic experiment where the objective is to study the effects of nutrients (factors) on the yield of corn. Even if only three major nutrients (nitrogen, phosphorus, and potash) and seven micronutrients (boron, copper, iron, magnesium, manganese, molybdenum, and zinc) are included in the experiment, each of these appearing at only two levels, namely presence or absence, there will be as many as $2^{10} = 1024$ treatment combinations. The use of a complete factorial will then involve at least 1024 experimental units and in most practical situations such a large experiment will be infeasible from considerations of cost and time. Similar examples abound in many other fields of application of factorial experiments.

Thus, unless the number of factors is sufficiently small, complete factorials are not affordable quite commonly in practice and one has no other alternative than including only some but not all treatment combinations in a factorial experiment. Obviously, such an experiment includes only a fraction of the totality of all possible treatment combinations and, therefore, the underlying experimental strategy is called a *fractional factorial plan*. Such a plan aims at drawing, under appropriate assumptions, valid statistical inference about the relevant factorial effects through an optimal utilization of the available resources. This entails challenging problems of both statistical and combinatorial interest.

The objective of the present book is to give an up-to-date and mathematical account of fractional factorial plans. For earlier accounts, up to various stages, we refer to Srivastava (1978), Raktoe, Hedayat, and Federer (1981), and Dey (1985). However, there has been a substantial growth of the subject since the publication of the last of these. Some of the basic ideas will be introduced in a more formal way in the next section. Then in Section 1.3 a brief outline of the contents of the subsequent chapters will be given.

1.2 PRELIMINARY IDEAS

Consider the setup of a factorial experiment involving $n(\geq 2)$ factors $F_1, \ldots,$ F_n, such that the ith factor F_i appears at m_i distinct levels, where $m_i \geq 2$, for $i = 1, \ldots, n$. An experiment of this kind is called an $m_1 \times \cdots \times m_n$ factorial experiment. If, in particular, $m_1 = \cdots = m_n = m$, say, then this setup corresponds to a symmetric m^n factorial; otherwise, it corresponds to an asymmetric or mixed factorial. For $1 \leq i \leq n$, let the levels of F_i be coded as $0, 1, \ldots, m_i - 1$. A typical treatment combination can then be represented by an n-tuple $j_1 \ldots j_n$, and the effect due to this treatment combination will be denoted by $\tau(j_1 \ldots j_n)$ $(0 \leq j_i \leq m_i - 1,\ 1 \leq i \leq n)$. There are altogether $v = \prod_{i=1}^{n} m_i$ treatment combinations that will hereafter be assumed to be lexicographically ordered. Let $\mathcal{V}$ denote the set of the v treatment combinations and $\boldsymbol{\tau}$ be a $v \times 1$ vector with elements $\tau(j_1 \ldots j_n)$ arranged in the lexicographic order. For example, if $n = 2, m_1 = 3, m_2 = 2$, then

$$\mathcal{V} = \{00,\ 01,\ 10,\ 11,\ 20,\ 21\}$$

and

$$\boldsymbol{\tau} = (\tau(00),\ \tau(01),\ \tau(10),\ \tau(11),\ \tau(20),\ \tau(21))'.$$

Similarly, if $n = 3$, $m_1 = m_2 = 2$, $m_3 = 3$, then $\boldsymbol{\tau} = (\tau(000), \tau(001),$ $\tau(002), \tau(010), \tau(011), \tau(012), \tau(100), \tau(101), \tau(102), \tau(110), \tau(111),$ $\tau(112))'$.

The treatment effects, that is, the elements of $\boldsymbol{\tau}$, are unknown parameters. A linear parametric function

$$\sum \cdots \sum l(j_1 \ldots j_n)\tau(j_1 \ldots j_n), \tag{1.2.1}$$

where $\{l(j_1 \ldots j_n)\}$ are real numbers, not all zeros, such that $\sum \cdots \sum l(j_1 \ldots j_n) = 0$ and the summation extends over $j_1 \ldots j_n \in \mathcal{V}$, is called a *treatment contrast*. In the context of factorial experiments, attention is focused on special types of treatment contrasts, namely those belonging to factorial effects. Following Bose (1947), a treatment contrast of the type (1.2.1) is said to belong to the factorial effect $F_{i_1} \ldots F_{i_g}$ $(1 \le i_1 < \cdots < i_g \le n;\ 1 \le g \le n)$ if

(i) $l(j_1 \ldots j_n)$ depends only on $j_{i_1}, \ldots, j_{i_g}$, and

(ii) writing $l(j_1 \ldots j_n) = \bar{l}(j_{i_1} \ldots j_{i_g})$ in consideration of (i) above, the sum of $\bar{l}(j_{i_1} \ldots j_{i_g})$ separately over each of the arguments $j_{i_1}, \ldots, j_{i_g}$ is zero.

By (i) and (ii), there are $\prod_{u=1}^{g}(m_{i_u} - 1)$ linearly independent contrasts belonging to the factorial effect $F_{i_1} \ldots F_{i_g}$. Clearly, there are $(2^n - 1)$ factorial effects in all. By the order of a factorial effect, we will mean the number of factors involved therein. A factorial effect is called a *main effect* if it involves exactly one factor (i.e., $g = 1$) and an *interaction* if it involves more than one factor (i.e., $g > 1$). Throughout, we will use the same notation for a factor and its main effect; namely, F_i will denote the ith factor as well as the main effect of the ith factor. We illustrate these ideas through an example.

Example 1.2.1. Consider the setup of a 3×2 factorial experiment. Then $n = 2$, $m_1 = 3$, $m_2 = 2$. By (1.2.1), (i) and (ii) above, contrasts belonging to main effects F_1, F_2 and the interaction F_1F_2 will, respectively, be of the forms

(a) $\sum_{j_1=0}^{2} \sum_{j_2=0}^{1} \bar{l}(j_1)\tau(j_1 j_2)$, where $\bar{l}(0), \bar{l}(1), \bar{l}(2)$ are not all zeros and

$$\bar{l}(0) + \bar{l}(1) + \bar{l}(2) = 0;$$

(b) $\sum_{j_1=0}^{2} \sum_{j_2=0}^{1} \bar{l}(j_2)\tau(j_1 j_2)$, where $\bar{l}(0)$ and $\bar{l}(1)$ are not both zeros and

$$\bar{l}(0) + \bar{l}(1) = 0;$$

(c) $\sum_{j_1=0}^{2} \sum_{j_2=0}^{1} \bar{l}(j_1 j_2)\tau(j_1 j_2)$, where $\{\bar{l}(j_1 j_2)\}$ are not all zeros and

$$\bar{l}(00) + \bar{l}(10) + \bar{l}(20) = \bar{l}(01) + \bar{l}(11) + \bar{l}(21) = 0,$$
$$\bar{l}(00) + \bar{l}(01) = \bar{l}(10) + \bar{l}(11) = \bar{l}(20) + \bar{l}(21) = 0.$$

There are 2, 1, and 2 linearly independent contrasts belonging to main effects F_1, F_2 and interaction F_1F_2, respectively. More specifically, taking $(\bar{l}(0), \bar{l}(1), \bar{l}(2)) = (\frac{1}{2}, 0, -\frac{1}{2})$ and $(\frac{1}{2}, -1, \frac{1}{2})$ in (a),

$$L_1^{(1)} = \tfrac{1}{2}[\tau(00) + \tau(01) - \tau(20) - \tau(21)],$$

and

$$L_2^{(1)} = \tfrac{1}{2}[\tau(00) + \tau(01)] - \tau(10) - \tau(11) + \tfrac{1}{2}[\tau(20) + \tau(21)]$$

represent two linearly independent contrasts belonging to the main effect F_1. Similarly, by (b),

$$L^{(2)} = \tfrac{1}{3}[\tau(00) + \tau(10) + \tau(20) - \tau(01) - \tau(11) - \tau(21)]$$

represents a contrast belonging to the main effect F_2. Finally, by (c),

$$L_1^{(12)} = \tau(00) - \tau(01) - \tau(20) + \tau(21)$$

and

$$L_2^{(12)} = \tau(00) - \tau(01) - 2\tau(10) + 2\tau(11) + \tau(20) - \tau(21)$$

are two linearly independent contrasts belonging to the interaction F_1F_2. It is not hard to interpret these contrasts. Thus $L_1^{(1)}$, $L_2^{(1)}$ measure the consequence of a change in the level of F_1 averaged over the levels of F_2; $L^{(2)}$ has a similar interpretation with roles of F_1, F_2 interchanged and $L_1^{(12)}$, $L_2^{(12)}$ measure the extent to which the consequence of a change in the level of one factor depends on the level of the other factor. ■

A few more definitions will be useful in the sequel. By (1.2.1), a typical treatment contrast is of the form $\boldsymbol{l}'\boldsymbol{\tau}$, where the $v \times 1$ vector $\boldsymbol{l}$ is nonnull and the sum of elements of $\boldsymbol{l}$ equals zero. A contrast $\boldsymbol{l}'\boldsymbol{\tau}$ is said to be *normalized* if $\boldsymbol{l}'\boldsymbol{l} = 1$. Several treatment contrasts are comparable when they are all normalized, that is, properly scaled. Hereafter, we will often consider normalized contrasts. Two treatment contrasts $\boldsymbol{l}'\boldsymbol{\tau}$ and $\boldsymbol{m}'\boldsymbol{\tau}$ are said to be *mutually orthogonal* if $\boldsymbol{l}'\boldsymbol{m} = 0$. One can easily verify that any two contrasts in Example 1.2.1 are mutually orthogonal (see Lemma 2.2.2, Chapter 2, for a general result in this regard). A set of treatment contrasts is called *orthonormal* if the contrasts in the set are normalized and mutually orthogonal.

With reference to an $m_1 \times \cdots \times m_n$ factorial setup, a *design* or *plan*, say d, is a rule that specifies the number of observations to be made with each treatment combination for inferring on parametric functions of interest. A plan d involving a totality of $N(> 0)$ observations is known as an N-observation or N-run plan. If $N < v$, the total number of treatment combinations, then d is called a *fractional factorial plan*. As indicated in the last section, fractional factorial plans are often indispensable in factorial experimentation and, as such, only plans of this kind will be considered in the sequel.

A plan d will be represented by listing all the treatment combinations such that each of them is repeated as many times as it arises in d; unless otherwise specified, we will have no reason to be concerned with the ordering within the list. For example, in a 3×2 factorial, with treatment combinations 00, 01, 10, 11, 20, 21, a plan that specifies two observations with 00, one observation with each of 01, 10, 20 and no observation with 11 or 21 will be exhibited as $\{00,\ 00,\ 01,\ 10,\ 20\}$.

1.3 SCOPE OF THE BOOK

The purpose of this book is to present a unified and up-to-date treatment of the mathematical theory of fractional factorial plans. The emphasis will be on the issues of optimality and construction. A background in matrix algebra and linear statistical models (see Rao, 1973a, chs. 1 and 4) will help the reader. Some familiarity with the general area of experimental design will also be advantageous, and in this connection, reference may be made to Cox (1958), Chakrabarti (1962), Raghavarao (1971), Dey (1986), Street and Street (1987), or Hinkelmann and Kempthorne (1994). Our treatment of fractional factorial plans relies heavily on a Kronecker calculus to be introduced in the next chapter. We refer to Gupta and Mukerjee (1989) for a review of this calculus in the context of complete factorials. A brief review is available also in Chatterjee (1982).

The mathematical foundation of our approach to fractional factorial plans is laid down in Chapter 2 where we systematically explore how, under the absence of higher-order factorial effects, such plans can allow the experimenter to draw valid inference on the lower-order ones that are typically the objects of interest in a factorial setting. The notion of optimality, which is the central theme of this book, will be introduced in the same chapter. So will be the concept of orthogonal arrays which play a key role in this connection. We close Chapter 2 by showing that plans based on orthogonal arrays are optimal in a very strong sense.

Because of the intimate link between orthogonal arrays and optimal fractional factorial plans, we find it appropriate to discuss the issues of construction and existence of these arrays in some detail. This will be taken up in Chapters 3, 4, and 5. While the material presented in these chapters should be of both statistical and combinatorial interest, our coverage of orthogonal arrays has been only to the extent to which they are relevant in the context of fractional factorial plans. Incidentally, Chapter 4 is an expanded version of an earlier review article by the authors (Dey and Mukerjee, 1998).

Although orthogonal arrays are of great help in the quest for optimal plans, their existence is not guaranteed in all experimental situations. The study of optimal fractional factorial plans, in the absence of orthogonal arrays, has received considerable attention in the literature. In Chapter 6 we review these results and some other related developments.

Chapter 7 addresses the issue of optimality when the experiment is conducted over a period of time (as happens commonly in the physical and engineering sciences) and the successive observations are influenced by a time trend in addition to the effects attributable to the factors. The related problem of minimizing the cost of factor level changes is also briefly discussed in this chapter. Finally, in the same chapter, we consider situations, where the experimental units are not homogeneous but can be grouped into homogeneous classes known as blocks, and explore the relevant fractional factorial plans.

The final chapter, Chapter 8, differs in spirit from the earlier ones in the sense of allowing a greater flexibility in the underlying linear model. This chapter starts with a discussion on the classical regular fractions and then briefly introduces certain newly emerging areas such as minimum aberration designs and also search and supersaturated designs.

As will be seen later in the book, Hadamard matrices and difference matrices are of much importance in connection with orthogonal arrays and fractional factorial plans. The book concludes with an appendix where we indicate the construction of Hadamard matrices of orders less than 50 and present some small difference matrices. Finally, as a ready reference for the reader, we provide in the appendix a guide to the construction of selected orthogonal arrays having up to 50 rows.

CHAPTER 2

Fractional Plans and Orthogonal Arrays

2.1 INTRODUCTION

The purpose of this chapter is to develop a theory of fractional factorials, based on a calculus for factorial arrangements. To begin with, in Section 2.2 we introduce the Kronecker notation, which is crucial in this context. By way of Kronecker calculus, in Section 2.3 a mathematical foundation is laid for our approach to the theory of fractional factorials. The important notion of resolution of a fractional factorial plan is introduced in Section 2.4 and some preliminaries needed for the exploration of optimal plans are presented in Section 2.5. Finally in Section 2.6 orthogonal arrays are introduced and the connection of these arrays with optimal fractional factorial plans discussed.

2.2 KRONECKER NOTATION

The use of Kronecker product of matrices (e.g., see Rao, 1973a, ch. 1) helps in expressing our notation in a very compact and convenient form. This approach was introduced formally by Kurkjian and Zelen (1962, 1963), although it seems that some of their ideas were inherent in the earlier works of Zelen (1958) and Shah (1958). Recall that if $A = (a_{ij})$ and $B = (b_{ij})$ are matrices of orders $m \times n$ and $p \times q$, respectively, then the Kronecker product of A and B, denoted by $A \otimes B$, is an $mp \times nq$ matrix C where $C = (a_{ij}B)$. The following are some well known and easily verifiable properties of Kronecker product of matrices:

(a) $(A \otimes B)(C \otimes D) = AC \otimes BD$ provided that AC and BD are well defined.

(b) $(A_1 + A_2) \otimes B = A_1 \otimes B + A_2 \otimes B$.

(c) $A \otimes (B_1 + B_2) = A \otimes B_1 + A \otimes B_2$.

(d) $aA \otimes bB = ab(A \otimes B)$, where a, b are scalars.

(e) $(A \otimes B)' = A' \otimes B'$.

(f) If A and B are invertible matrices, then $(A \otimes B)^{-1} = A^{-1} \otimes B^{-1}$.

We now consider the setup of an $m_1 \times \cdots \times m_n$ factorial experiment involving $n (\geq 2)$ factors $F_1, \ldots, F_n$ appearing at $m_1, \ldots, m_n (\geq 2)$ levels, respectively, and introduce the essentials of Kronecker notation. As in Section 1.2, we write $v = \prod_{i=1}^{n} m_i$ and denote the $v \times 1$ vector, with elements $\tau(j_1 \ldots j_n)$ arranged in the lexicographic order, by $\boldsymbol{\tau}$. Let Ω be the set of all binary n-tuples and Ω^* be the subset of Ω consisting of its nonnull members. For each $\boldsymbol{x} = x_1 x_2 \ldots x_n \in \Omega$, define

$$\alpha(\boldsymbol{x}) = \prod_{i=1}^{n} (m_i - 1)^{x_i}. \tag{2.2.1}$$

It is easy to see that there is a one–one correspondence between Ω^* and the set of all factorial effects in the sense that a typical factorial effect $F_{i_1} \ldots F_{i_g} (1 \leq i_1 < \cdots < i_g \leq n; 1 \leq g \leq n)$ corresponds to the element $\boldsymbol{x} = x_1 x_2 \ldots x_n$ of Ω^* such that $x_{i_1} = \cdots = x_{i_g} = 1$ and $x_u = 0$ for $u \neq i_1, \ldots, i_g$. Thus the $2^n - 1$ factorial effects may be represented by $F^{\boldsymbol{x}}, \boldsymbol{x} \in \Omega^*$. For example, if $n = 2$, then $\Omega = \{00, 01, 10, 11\}$, $\Omega^* = \{01, 10, 11\}$ and main effects F_1, F_2, and the two-factor interaction $F_1 F_2$ may be represented as F^{10}, F^{01}, and F^{11}, respectively. As noted in Section 1.2, for each $\boldsymbol{x} \in \Omega^*$, there are $\alpha(\boldsymbol{x})$ linearly independent treatment contrasts belonging to $F^{\boldsymbol{x}}$ where $\alpha(\boldsymbol{x})$ is given by (2.2.1).

For a positive integer ν, let $\mathbf{1}_\nu$ be the $\nu \times 1$ vector with each element unity and I_ν be the identity matrix of order ν. For $1 \leq i \leq n$, let P_i be an $(m_i - 1) \times m_i$ matrix such that the $m_i \times m_i$ matrix $(m_i^{-1/2} \mathbf{1}_{m_i}, P_i')$ is orthogonal, namely, such that

$$P_i \mathbf{1}_{m_i} = \mathbf{0}, \quad P_i P_i' = I_{m_i - 1}. \tag{2.2.2}$$

Thus, if $n = 2$, $m_1 = 3$, $m_2 = 2$, then we can take

$$P_1 = \begin{pmatrix} \frac{1}{\sqrt{2}} & 0 & \frac{-1}{\sqrt{2}} \\ \frac{1}{\sqrt{6}} & \frac{-2}{\sqrt{6}} & \frac{1}{\sqrt{6}} \end{pmatrix}, \quad P_2 = \begin{pmatrix} \frac{1}{\sqrt{2}} & \frac{-1}{\sqrt{2}} \end{pmatrix}. \tag{2.2.3}$$

While P_i satisfying (2.2.2) is nonunique, it may be emphasized that our main ideas and conclusions *do not* depend on the specific choice of P_i, $1 \leq i \leq n$; see Remark 2.3.1 below for more details. For each $\boldsymbol{x} = x_1 \ldots x_n \in \Omega$, define the $\alpha(\boldsymbol{x}) \times v$ matrix

$$P^{\boldsymbol{x}} = P_1^{x_1} \otimes \cdots \otimes P_n^{x_n} = \bigotimes_{i=1}^{n} P_i^{x_i}, \tag{2.2.4}$$

where, as before, $\otimes$ denotes the Kronecker product of matrices and for $1 \leq i \leq n$,

$$P_i^{x_i} = \begin{cases} m_i^{-1/2}\mathbf{1}'_{m_i} & \text{if } x_i = 0, \\ P_i & \text{if } x_i = 1. \end{cases} \tag{2.2.5}$$

We now prove the following result:

Lemma 2.2.1. *For each* $\boldsymbol{x}, \boldsymbol{y} \in \Omega$, $\boldsymbol{x} \neq \boldsymbol{y}$,

(a) $P^{\boldsymbol{x}}(P^{\boldsymbol{x}})' = I_{\alpha(\boldsymbol{x})}$,
(b) $P^{\boldsymbol{x}}(P^{\boldsymbol{y}})' = O$, *where O is a null matrix.*

Proof. The statement (a) is obvious from (2.2.1), (2.2.2), (2.2.5), and the definition of $P^{\boldsymbol{x}}$. To prove (b), note that from (2.2.4),

$$P^{\boldsymbol{x}}(P^{\boldsymbol{y}})' = \bigotimes_{i=1}^{n} P_i^{x_i}(P_i^{y_i})'.$$

Now $\boldsymbol{x} \neq \boldsymbol{y}$ implies that $x_i \neq y_i$ for some i and for this i, by (2.2.2) and (2.2.5), $P_i^{x_i}(P_i^{y_i})'$ equals the null row or column vector. This proves the result. ■

From (2.2.4) and (2.2.5), it is easy to observe that for each $\boldsymbol{x} \in \Omega^*$, the elements of $P^{\boldsymbol{x}}\boldsymbol{\tau}$ are treatment contrasts belonging to $F^{\boldsymbol{x}}$. Also by Lemma 2.2.1(a), Rank $(P^{\boldsymbol{x}})(= \alpha(\boldsymbol{x}))$ equals the number of linearly independent treatment contrasts belonging to $F^{\boldsymbol{x}}$. Hence the following result, giving a convenient representation for contrasts belonging to factorial effects, is evident from Lemma 2.2.1:

Lemma 2.2.2. *For each $\boldsymbol{x} \in \Omega^*$, the elements of $P^{\boldsymbol{x}}\boldsymbol{\tau}$ represent a complete set of orthonormal treatment contrasts belonging to the factorial effect $F^{\boldsymbol{x}}$. Furthermore contrasts belonging to different factorial effects are mutually orthogonal.* ■

It is also possible to provide an interpretation for $P^{00\ldots0}\boldsymbol{\tau}$. By (2.2.4) and (2.2.5), $P^{00\ldots0} = v^{-1/2}\mathbf{1}'_v$ and hence

$$P^{00\ldots0}\boldsymbol{\tau} = v^{1/2}\bar{\tau}, \tag{2.2.6}$$

where $\bar{\tau}$, the general mean, is the arithmetic mean of the $\{\tau(j_1\ldots j_n)\}$.

Example 2.2.1. Continuing with the setup of Example 1.2.1, let P_1 and P_2 be chosen as in (2.2.3). One can find P^{10} explicitly using (2.2.4) and (2.2.5) and check that the elements of $P^{10}\boldsymbol{\tau}$ are normalized versions of $L_1^{(1)}$ and $L_2^{(1)}$ as introduced in Example 1.2.1. Similarly it may be seen that $P^{01}\boldsymbol{\tau}$ and $P^{11}\boldsymbol{\tau}$ represent normalized versions of $L^{(2)}$ and $L_1^{(12)}$, $L_2^{(12)}$, respectively. ■

The following lemma will be required in the next section:

Lemma 2.2.3. *For each $\boldsymbol{x} \in \Omega$, no column of $P^{\boldsymbol{x}}$ can equal the null vector.*

Proof. If $P^{\boldsymbol{x}}$ has a null column for some $\boldsymbol{x} \in \Omega$, then by (2.2.4), (2.2.5), P_i must have a null column for some i. This, however, is impossible since for each i the matrix $(m_i^{-1/2}\mathbf{1}_{m_i}, P_i')'$ is orthogonal, so each column of this matrix must have unit norm. ■

Before concluding this section, we briefly discuss a choice of P_i which is often used in the context of a quantitative factor. Without loss of generality, let F_1 be such a factor with $\kappa_0, \ldots, \kappa_{m_1-1}$ representing distinct quantitative "doses," measured in an appropriate unit, corresponding to the levels $0, \ldots, m_1 - 1$, respectively. Let $q_u(\cdot)$, $1 \le u \le m_1 - 1$, be orthogonal polynomials such that $q_u(\cdot)$ is of degree u and

$$\sum_{j=0}^{m_1-1} q_u(\kappa_j) = 0 \quad (1 \le u \le m_1 - 1),$$

$$\sum_{j=0}^{m_1-1} q_u(\kappa_j)q_{u'}(\kappa_j) = \delta_{uu'} \quad (1 \le u, u' \le m_1 - 1),$$

where $\delta_{uu'}$ is the Kronecker delta, which equals 1 if $u = u'$ and 0 otherwise. Then the $(m_1 - 1) \times m_1$ matrix

$$P_1 = ((q_u(\kappa_j))), \quad 1 \le u \le m_1 - 1, 0 \le j \le m_1 - 1, \tag{2.2.7}$$

satisfies (2.2.2). For example, if $m_1 = 3$ and $\kappa_1 = (\kappa_0 + \kappa_2)/2$ (i.e., the doses are equispaced), then one can take

$$q_1(\rho) = \sqrt{2}\,\frac{\kappa_1 - \rho}{\kappa_2 - \kappa_0},$$

$$q_2(\rho) = \sqrt{24}\left[\left(\frac{\kappa_1 - \rho}{\kappa_2 - \kappa_0}\right)^2 - \frac{1}{6}\right],$$

and satisfy oneself that (2.2.7) reduces to the expression for P_1 given in (2.2.3). With P_1 constructed as in (2.2.7) using orthogonal polynomials, the successive elements of $P^{10\ldots0}\boldsymbol{\tau}$ can be interpreted as contrasts belonging to main effect F_1 and representing its linear, quadratic, and other components. In general, if there are several quantitative factors, then choosing the corresponding P_i's as in (2.2.7), it is possible to interpret contrasts belonging to interactions among these factors, as given by the elements of the corresponding $P^x\boldsymbol{\tau}$'s, in a similar way. We do not elaborate on this point here, but refer the interested reader to Raktoe, Hedayat, and Federer (1981) for a fuller account.

2.3 FRACTIONAL FACTORIAL PLANS

Consider now an N-run fractional factorial plan d for an $m_1 \times \cdots \times m_n$ factorial, where $0 < N < v(= \prod m_i)$. According to the plan d, suppose that $r_d(j_1 \ldots j_n)$ observations are to be made with treatment combination $j_1 \ldots j_n$ for each $j_1 \ldots j_n \in \mathcal{V}$, the set of all the v treatment combinations, where $\{r_d(j_1 \ldots j_n)\}$ are nonnegative integers satisfying

$$\sum_{j_1 \ldots j_n \in \mathcal{V}} \cdots \sum r_d(j_1 \ldots j_n) = N. \tag{2.3.1}$$

For $1 \le u \le N$, let Y_u be the uth observation according to d. If Y_u corresponds to the treatment combination $j_1 \ldots j_n$, then we will assume that $\mathbb{E}(Y_u)$ equals $\tau(j_1 \ldots j_n)$, where $\mathbb{E}(\cdot)$ stands for expectation. Furthermore $Y_1, \ldots, Y_N$ are assumed to be uncorrelated and to have a common variance $\sigma^2(> 0)$.

It will be convenient to express the above model in matrix notation. To that end, for $1 \le u \le N$, $j_1 \ldots j_n \in \mathcal{V}$, define the indicator $\chi_d(u; j_1 \ldots j_n)$, which assumes the value 1 if the uth observation according to d corresponds to $j_1 \ldots j_n$ and the value zero otherwise. Let X_d be the $N \times v$ design matrix, with rows indexed by u and columns indexed by $j_1 \ldots j_n$, such that

$$X_d = (\chi_d(u; j_1 \ldots j_n))_{\substack{u=1,\ldots,N \\ j_1 \ldots j_n \in \mathcal{V}}}. \tag{2.3.2}$$

The columns of X_d are assumed to be lexicographically ordered. Then, with $\boldsymbol{Y} = (Y_1, \ldots, Y_N)'$, the linear model can be expressed as

$$\mathbb{E}(\boldsymbol{Y}) = X_d\boldsymbol{\tau}, \quad \mathbb{D}(\boldsymbol{Y}) = \sigma^2 I_N, \tag{2.3.3}$$

where $\mathbb{E}$, as before, stands for the expectation and $\mathbb{D}$ for the dispersion (variance-covariance) matrix.

A linear parametric function of the form $\boldsymbol{l}'\boldsymbol{\tau}$ will be said to be estimable in d if it has an unbiased estimator which is a linear function of $\boldsymbol{Y}$. By (2.3.3), this happens if and only if $\boldsymbol{l}' \in \mathcal{R}(X_d)$, where $\mathcal{R}(\cdot)$ denotes the row space of a matrix. Hence for any $\boldsymbol{x} \in \Omega$, $P^{\boldsymbol{x}}\boldsymbol{\tau}$ is estimable in d (i.e., each element of $P^{\boldsymbol{x}}\boldsymbol{\tau}$ is estimable in d) if and only if

$$\mathcal{R}(P^{\boldsymbol{x}}) \subset \mathcal{R}(X_d). \tag{2.3.4}$$

Now if $r_d(j_1 \ldots j_n) = 0$ for some $j_1 \ldots j_n \in \mathcal{V}$, then the corresponding column of X_d equals the null vector and hence by Lemma 2.2.3, the relation (2.3.4) cannot hold for any $\boldsymbol{x} \in \Omega$. In other words, for any $\boldsymbol{x} \in \Omega$, in order to ensure the estimability of $P^{\boldsymbol{x}}\boldsymbol{\tau}$ in d, it is necessary that $r_d(j_1 \ldots j_n) \geq 1$ for each $j_1 \ldots j_n \in \mathcal{V}$, namely, $N \geq v$, by (2.3.1). Because of Lemma 2.2.2, this means in particular that unless further assumptions are made, a fractional factorial plan is incapable even of ensuring the estimability of complete sets of main effect contrasts which are invariably the parametric functions of interest in any factorial setup.

It is possible to overcome this difficulty if, from a knowledge of the physical process underlying the experimental setup or from past experience, one can assume the absence or negligibility of higher-order effects. Fortunately such an assumption is reasonable in many practical situations. Suppose that one can assume the absence of factorial effects involving $t+1$ or more factors, where $1 \leq t \leq n-1$. Then by Lemma 2.2.2,

$$P^{\boldsymbol{x}}\boldsymbol{\tau} = 0 \quad \text{for each } \boldsymbol{x} \in \bar{\Omega}_t, \tag{2.3.5}$$

where $\bar{\Omega}_t$ is a subset of Ω consisting of those binary n-tuples that have at least $t+1$ components equal to unity. Define $P = (\ldots, (P^{\boldsymbol{x}})', \ldots)'_{\boldsymbol{x} \in \Omega}$. For example, if $n = 2$, then

$$P = \begin{bmatrix} P^{00} \\ P^{01} \\ P^{10} \\ P^{11} \end{bmatrix}.$$

The matrix P has $\sum_{x \in \Omega} \alpha(\boldsymbol{x})$ rows and v columns. But by virtue of (2.2.1), $\sum_{x \in \Omega} \alpha(\boldsymbol{x}) = v$. Hence by Lemma 2.2.1, P is an orthogonal matrix of order v. Therefore, under (2.3.5), $\boldsymbol{\tau}$ can be expressed as

$$\boldsymbol{\tau} = \sum_{x \in \Omega_t} (P^x)' \boldsymbol{\beta}_x, \tag{2.3.6}$$

where $\Omega_t = \Omega - \bar{\Omega}_t$ (i.e., Ω_t consists of binary n-tuples with at most t components unity) and for each $\boldsymbol{x} \in \Omega_t$, $\boldsymbol{\beta}_x$ is some $\alpha(\boldsymbol{x}) \times 1$ vector. Under (2.3.5), by Lemma 2.2.1 and (2.3.6),

$$\boldsymbol{\beta}_x = P^x \boldsymbol{\tau}, \quad \text{for each } \boldsymbol{x} \in \Omega_t, \tag{2.3.7}$$

which provides an interpretation for $\boldsymbol{\beta}_x$ $(\boldsymbol{x} \in \Omega_t, \boldsymbol{x} \neq 00\ldots0)$ in terms of a complete set of orthonormal contrasts belonging to a possibly nonnegligible factorial effect. Also, by (2.2.6) and (2.3.7), $\beta_{00\ldots0}$ can be interpreted in terms of the general mean.

Under the absence of factorial effects involving $t + 1$ or more factors, suppose that interest lies in the general mean and contrasts belonging to the lower-order factorial effects, say those involving at most f factors, where $1 \leq f \leq t$. In other words, recalling (2.2.6) and Lemma 2.2.2, we consider a situation where, under (2.3.5), one is interested in $P^x \boldsymbol{\tau}$ for $\boldsymbol{x} \in \Omega_f$, where $1 \leq f \leq t$. Equivalently, by (2.3.7), the objects of interest are then $\boldsymbol{\beta}_x$ for $\boldsymbol{x} \in \Omega_f (\subset \Omega_t)$; namely interest lies in

$$\boldsymbol{\beta}^{(1)} = (\ldots, \boldsymbol{\beta}_x{}', \ldots)'_{x \in \Omega_f}. \tag{2.3.8}$$

Define

$$P^{(1)} = (\ldots, (P^x)', \ldots)'_{x \in \Omega_f}, \tag{2.3.9}$$

and if $f < t$, in which case Ω_f is a proper subset of Ω_t, let

$$\boldsymbol{\beta}^{(2)} = (\ldots, \boldsymbol{\beta}_x{}', \ldots)'_{x \in \Omega_t - \Omega_f}, \; P^{(2)} = (\ldots, (P^x)', \ldots)'_{x \in \Omega_t - \Omega_f}. \tag{2.3.10}$$

For example, if $n = 3$, $f = 1, t = 2$, then

$$\Omega_f = \{000, 100, 010, 001\}, \;\; \Omega_t - \Omega_f = \{011, 101, 110\},$$
$$\boldsymbol{\beta}^{(1)} = (\boldsymbol{\beta}_{000}{}', \boldsymbol{\beta}_{100}{}', \boldsymbol{\beta}_{010}{}', \boldsymbol{\beta}_{001}{}')', \;\; \boldsymbol{\beta}^{(2)} = (\boldsymbol{\beta}_{011}{}', \boldsymbol{\beta}_{101}{}', \boldsymbol{\beta}_{110}{}')',$$

and similarly $P^{(1)}$ and $P^{(2)}$ are defined. By (2.3.8) and (2.3.10), $\boldsymbol{\beta}^{(1)}$ and $\boldsymbol{\beta}^{(2)}$ are column vectors of orders α_f and $\alpha_t - \alpha_f$, respectively, where, for

$0 \le \nu \le n$,

$$\alpha_\nu = \sum_{x \in \Omega_\nu} \alpha(\boldsymbol{x}), \tag{2.3.11}$$

$\alpha(\boldsymbol{x})$ being as in (2.2.1). Similarly $P^{(1)}$ and $P^{(2)}$ are $\alpha_f \times \upsilon$ and $(\alpha_t - \alpha_f) \times \upsilon$ matrices, respectively.

Using (2.3.8)–(2.3.10), the relation (2.3.6) can be written as

$$\boldsymbol{\tau} = (P^{(1)})'\boldsymbol{\beta}^{(1)} + (P^{(2)})'\boldsymbol{\beta}^{(2)}.$$

Hence, under (2.3.5), the linear model (2.3.3), associated with an N-run fractional factorial plan d, is expressible as

$$\mathbb{E}(\boldsymbol{Y}) = (X_{1d} \vdots X_{2d}) \begin{pmatrix} \boldsymbol{\beta}^{(1)} \\ \boldsymbol{\beta}^{(2)} \end{pmatrix}, \quad \mathbb{D}(\boldsymbol{Y}) = \sigma^2 I_N \tag{2.3.12}$$

where

$$X_{1d} = X_d(P^{(1)})', \quad X_{2d} = X_d(P^{(2)})'. \tag{2.3.13}$$

For any matrix A, let A^- denote a generalized inverse of A, that is, $AA^-A = A$. Define the $\alpha_f \times \alpha_f$ matrix

$$\mathcal{I}_d = X_{1d}'X_{1d} - X_{1d}'X_{2d}(X_{2d}'X_{2d})^- X_{2d}'X_{1d}. \tag{2.3.14}$$

The second term in the expression for $\mathcal{I}_d$ does not arise if $f = t$. We are now in a position to present a result giving a necessary and sufficient condition for the estimability of $\boldsymbol{\beta}^{(1)}$ in d under the absence of factorial effects involving $t+1$ factors or more, namely under (2.3.12).

Theorem 2.3.1. *Under the model (2.3.12), $\boldsymbol{\beta}^{(1)}$ is estimable in d if and only if $\mathcal{I}_d$ is positive definite (p.d.). Further, if this condition holds, then $\mathbb{D}(\hat{\boldsymbol{\beta}}^{(1)}) = \sigma^2\mathcal{I}_d^{-1}$, where $\hat{\boldsymbol{\beta}}^{(1)} = \mathcal{I}_d^{-1}V_d'\boldsymbol{Y}$ is the best linear unbiased estimator (BLUE) of $\boldsymbol{\beta}^{(1)}$ arising from d.*

Proof. Note that

$$\mathcal{I}_d = V_d'V_d, \tag{2.3.15}$$

where

$$V_d = (I_N - X_{2d}(X_{2d}'X_{2d})^- X_{2d}')X_{1d}. \tag{2.3.16}$$

Under the model (2.3.12), if $\boldsymbol{\beta}^{(1)}$ is estimable in d, then there exists a matrix G_d of order $\alpha_f \times N$ such that $G_d(X_{1d} \vdots X_{2d}) = (I_{\alpha_f} \vdots O)$, where O is a null matrix of order $\alpha_f \times (\alpha_t - \alpha_f)$. Then by (2.3.16), $G_d V_d = I_{\alpha_f}$, so V_d has full column rank and hence by (2.3.15), $\mathcal{I}_d$ is p.d.

Conversely, if $\mathcal{I}_d$ is p.d., then by (2.3.14) and (2.3.16),

$$\mathcal{I}_d^{-1} V_d'(X_{1d} \vdots X_{2d}) = (I_{\alpha_f} \vdots O)$$

so that $\boldsymbol{\beta}^{(1)}$ is estimable in d under the model (2.3.12). Furthermore, noting that

$$\mathcal{R}(\mathcal{I}_d{}^{-1} V_d') \subset \mathcal{R}(V_d') \subset \mathcal{R}\begin{pmatrix} X_{1d}' \\ X_{2d}' \end{pmatrix},$$

the BLUE of $\boldsymbol{\beta}^{(1)}$ is then given by $\hat{\boldsymbol{\beta}}^{(1)} = \mathcal{I}_d{}^{-1} V_d' \boldsymbol{Y}$. Hence by (2.3.12) and (2.3.15), $\mathbb{D}(\hat{\boldsymbol{\beta}}^{(1)}) = \sigma^2 \mathcal{I}_d^{-1}$. ■

The matrix $\mathcal{I}_d$ is called the information matrix for $\boldsymbol{\beta}^{(1)}$ under the plan d and the model (2.3.12). As Theorem 2.3.1 shows, this matrix plays a crucial role when interest lies in drawing inference on $\boldsymbol{\beta}^{(1)}$; we will consider this point in greater detail in Section 2.5.

An alternative expression for $\mathcal{I}_d$ will be useful. Let R_d be a $v \times v$ diagonal matrix with diagonal elements $r_d(j_1 \ldots j_n)$ arranged lexicographically. By (2.3.2),

$$X_d' X_d = R_d.$$

Hence by (2.3.13), (2.3.14), $\mathcal{I}_d$ can be expressed as

$$\mathcal{I}_d = P^{(1)} R_d (P^{(1)})' - P^{(1)} R_d (P^{(2)})' [P^{(2)} R_d (P^{(2)})']^{-} P^{(2)} R_d (P^{(1)})', \tag{2.3.17}$$

where the second term does not arise if $f = t$.

We now return to the issue that led us to consider a setup where, as indicated by (2.3.5), higher-order factorial effects can be assumed to be absent. Under the model (2.3.12), resulting from (2.3.5), is it possible to obtain a fractional factorial plan d that keeps $\mathcal{I}_d$ p.d. and hence $\boldsymbol{\beta}^{(1)}$, the vector representing the parameters of interest, estimable? As the next example and, more formally, the discussion following the proof of Theorem 2.3.2 below reveal, the answer to this question is in the affirmative.

Example 2.3.1. Consider a 4-run fractional factorial plan

$$d = \{000, 011, 101, 110\}$$

for a 2^3 factorial. Let $f = t = 1$, that is, interest lies in the general mean and the main effect contrasts, under the assumption of absence of all interactions involving two or more factors. Since $f = t$, by (2.3.17), $\mathcal{I}_d = P^{(1)} R_d (P^{(1)})'$. Here $m_1 = m_2 = m_3 = 2$. Hence, by (2.2.2), $P_i = \pm(\frac{1}{\sqrt{2}}, \frac{-1}{\sqrt{2}})$, for each i. Therefore, by (2.2.4), (2.2.5), $P^{(1)}$ is a 4×8 matrix with rows

$$2^{-3/2}\{(1, 1) \otimes (1, 1) \otimes (1, 1)\},$$
$$\pm 2^{-3/2}\{(1, -1) \otimes (1, 1) \otimes (1, 1)\},$$
$$\pm 2^{-3/2}\{(1, 1) \otimes (1, -1) \otimes (1, 1)\},$$

and

$$\pm 2^{-3/2}\{(1, 1) \otimes (1, 1) \otimes (1, -1)\}.$$

Also $R_d = \text{diag}(1, 0, 0, 1, 0, 1, 1, 0)$. Hence $\mathcal{I}_d = \frac{1}{2} I_4$, and by Theorem 2.3.1, d ensures the estimability of $\boldsymbol{\beta}^{(1)}$ under the model (2.3.12). ■

More generally, given $n, m_1, \ldots, m_n, f$ and t, one may wish to know about the smallest possible value of N such that an N-run plan, that keeps $\boldsymbol{\beta}^{(1)}$ estimable under (2.3.12), is available. A simple solution is presented below for the case $f = t$.

Theorem 2.3.2. *Let $f = t$. Then an N-run plan, keeping $\boldsymbol{\beta}^{(1)}$ estimable under (2.3.12) is available if and only if $N \geq \alpha_f$, where α_f is given by (2.3.11).*

Proof. Suppose an N-run plan d, ensuring the estimability of $\boldsymbol{\beta}^{(1)}$ under (2.3.12), is available. Since $f = t$, by (2.3.17) and Theorem 2.3.1,

$$\alpha_f = \text{Rank}(\mathcal{I}_d) = \text{Rank}(P^{(1)} R_d (P^{(1)})') \leq \text{Rank}(R_d).$$

Recalling the definition of R_d, it follows that $r_d(j_1 \ldots j_n) \geq 1$ for at least α_f different choices of $j_1 \ldots j_n$. Hence by (2.3.1), $N \geq \alpha_f$. Conversely, let $N \geq \alpha_f$. For any plan d, by (2.3.2) and (2.3.13), each row of X_{1d} is also a row of $(P^{(1)})'$. But by (2.3.9) and Lemma 2.2.1, $P^{(1)}(P^{(1)})' = I_{\alpha_f}$; that is, the $v \times \alpha_f$ matrix $(P^{(1)})'$ has rank α_f. Hence if $N \geq \alpha_f$, then one can find an N-run plan d such that the $N \times \alpha_f$ matrix X_{1d} has full column rank; for this it is enough to choose d such that some α_f linearly independent rows of

$(P^{(1)})'$ are included as rows in X_{1d}. Then by (2.3.14) (note that here $f = t$), $\mathcal{I}_d = X'_{1d}X_{1d}$ is p.d. so that, by Theorem 2.3.1, d allows the estimability of $\boldsymbol{\beta}^{(1)}$ under (2.3.12). ■

Clearly $\alpha_f < v$ so that by Theorem 2.3.2, if $f = t$, then under the model (2.3.12) the existence of a fractional factorial plan that keeps $\boldsymbol{\beta}^{(1)}$ estimable is guaranteed. The same conclusion holds also for $f < t$ since then the same argument as with Theorem 2.3.2 shows the availability of an $\alpha_t(< v)$-run plan retaining the estimability of $(\boldsymbol{\beta}^{(1)'}, \boldsymbol{\beta}^{(2)'})'$ and hence that of $\boldsymbol{\beta}^{(1)}$. In fact, as will be evident later from Theorem 2.6.1 and, say, (2.6.3), under (2.3.12) with $f < t$ one may be able to keep $\boldsymbol{\beta}^{(1)}$ estimable using even fewer than α_t runs.

As Theorem 2.3.2 shows, if $f = t$, then the most economic plan, which keeps $\boldsymbol{\beta}^{(1)}$ estimable under (2.3.12), involves α_f runs. Such a plan is called *saturated*. The plan d of Example 2.3.1 is of this kind. We will discuss more about saturated plans in Chapter 6.

For $f = t$, while a saturated plan keeps $\boldsymbol{\beta}^{(1)}$ estimable, it leaves no degree of freedom for error and thus precludes the use of the standard theory (Rao, 1973a, ch. 4) for testing linear hypotheses concerning $\boldsymbol{\beta}^{(1)}$. Given a saturated plan, this necessitates the development of special techniques for testing hypotheses. These aspects are beyond the scope of this book but the reader may see Dong (1993), Kunert (1997), Voss (1997), Hamada and Balakrishnan (1998), and the references therein for details.

Remark 2.3.1. In the last section, it was mentioned that our main ideas and conclusions do not depend upon the specific choice of $\{P_1, \ldots, P_n\}$ satisfying (2.2.2). This point may be further elucidated at this stage. Consider any two distinct choices of $\{P_1, \ldots, P_n\}$, say, $\{P_{11}, \ldots, P_{1n}\}$ and $\{P_{21}, \ldots, P_{2n}\}$, both satisfying (2.2.2). Then it is easily seen that there exist orthogonal matrices $\Gamma_1, \ldots, \Gamma_n$ such that

$$P_{1i} = \Gamma_i P_{2i}, \quad 1 \le i \le n. \tag{2.3.18}$$

As a consequence, for a given plan d, if $\mathcal{I}_{1d}$ and $\mathcal{I}_{2d}$ stand for the expressions of $\mathcal{I}_d$ as obtained by starting from the choices $\{P_{11}, \ldots, P_{1n}\}$ and $\{P_{21}, \ldots, P_{2n}\}$, respectively, then from (2.3.17) and the definitions of $P^{(1)}$ and $P^{(2)}$ it follows that there exists an orthogonal matrix Γ such that

$$\mathcal{I}_{1d} = \Gamma \mathcal{I}_{2d} \Gamma'. \tag{2.3.19}$$

This shows that if $\mathcal{I}_d$ is p.d. for some choice of $\{P_1, \ldots, P_n\}$, then it remains so for every choice of $\{P_1, \ldots, P_n\}$ satisfying (2.2.2). Thus by Theorem 2.3.1, the status of a plan with regard to the estimability of the general mean and complete sets of orthonormal contrasts belonging to lower-order factorial effects, under the absence of higher-order factorial effects, is not influenced by the specific choice of $\{P_1, \ldots, P_n\}$. Even without Theorem 2.3.1, this is intuitively expected, since by (2.3.18), a change in $\{P_1, \ldots, P_n\}$ subject to (2.2.2), induces an orthogonal transformation on $\boldsymbol{\beta}^{(1)}$ and does not effectively alter (2.3.5).

By (2.3.19), the choice of $\{P_1, \ldots, P_n\}$ does not affect the eigenvalues of $\mathcal{I}_d$. Similarly, if for some choice of $\{P_1, \ldots, P_n\}$, $\mathcal{I}_d$ equals cI_{α_f}, where c is some positive constant, then for every choice of $\{P_1, \ldots, P_n\}$, $\mathcal{I}_d$ equals cI_{α_f}. Consequently the results on optimality, presented later in this chapter and also in subsequent chapters, do not depend on the specific choice of $\{P_1, \ldots, P_n\}$. ■

2.4 CONCEPT OF RESOLUTION

In the last section we studied fractional factorial plans that ensure the estimability of the general mean and complete sets of contrasts belonging to factorial effects involving at most f factors under the absence of factorial effects involving $t+1$ or more factors, where $1 \leq f \leq t \leq n-1$. Plans of this kind will hereafter be referred to as Resolution (f, t) plans. Thus by Theorem 2.3.1, d is a Resolution (f, t) plan if and only if the matrix $\mathcal{I}_d$, given by (2.3.17) with $P^{(1)}$ and $P^{(2)}$ as in (2.3.9) and (2.3.10), is p.d.

It is important to note here that in our definition of resolution of a fractional factorial plan, we are deviating from the more standard terminology (e.g., see Box and Hunter, 1961a, b), according to which Resolution (f, t) plans might have been referred to simply as Resolution $(f+t+1)$ plans. We attempt to provide an explanation. We could have followed the standard terminology had the status of a plan, with regard to the estimation problem considered in the last section (*vide* Theorem 2.3.1), depended on f and t exclusively through the sum $f+t$. While this indeed happens with plans based on orthogonal arrays discussed in Section 2.6, this may not be the case in general (it may be noted that the standard terminology had originally arisen in the context of plans based on orthogonal arrays). In other words, as illustrated by Example 2.4.1 below, a plan d may be of Resolution (f, t) but not of Resolution (f', t') and yet $f+t$ may equal $f'+t'$. We believe that this provides a justification for our deviation from the standard terminology; see also Webb (1968) in this connection.

Example 2.4.1. With reference to a 2^4 factorial, consider a 11-run fractional factorial plan

$$d = \{0000, 0001, 0010, 0011, 0100, 0101, 0110, 1000, 1001, 1010, 1100\}.$$

Then, $R_d = \text{diag}(1, 1, 1, 1, 1, 1, 1, 0, 1, 1, 1, 0, 1, 0, 0, 0)$. We will explore the behavior of d with $(f, t) = (2, 2)$ and $(1, 3)$; note that $f + t = 4$ in both cases. Let $\epsilon_1 = (1, 1)'$, $\epsilon_2 = (1, -1)'$, and $\Lambda_1, \Lambda_2, \Lambda_3$ be 5×16, 6×16 and 4×16 matrices defined as

$$\Lambda_1 = \left(\frac{1}{4}\right) \begin{bmatrix} \epsilon_1' \otimes \epsilon_1' \otimes \epsilon_1' \otimes \epsilon_1' \\ \epsilon_2' \otimes \epsilon_1' \otimes \epsilon_1' \otimes \epsilon_1' \\ \epsilon_1' \otimes \epsilon_2' \otimes \epsilon_1' \otimes \epsilon_1' \\ \epsilon_1' \otimes \epsilon_1' \otimes \epsilon_2' \otimes \epsilon_1' \\ \epsilon_1' \otimes \epsilon_1' \otimes \epsilon_1' \otimes \epsilon_2' \end{bmatrix},$$

$$\Lambda_2 = \left(\frac{1}{4}\right) \begin{bmatrix} \epsilon_1' \otimes \epsilon_1' \otimes \epsilon_2' \otimes \epsilon_2' \\ \epsilon_1' \otimes \epsilon_2' \otimes \epsilon_1' \otimes \epsilon_2' \\ \epsilon_1' \otimes \epsilon_2' \otimes \epsilon_2' \otimes \epsilon_1' \\ \epsilon_2' \otimes \epsilon_1' \otimes \epsilon_1' \otimes \epsilon_2' \\ \epsilon_2' \otimes \epsilon_1' \otimes \epsilon_2' \otimes \epsilon_1' \\ \epsilon_2' \otimes \epsilon_2' \otimes \epsilon_1' \otimes \epsilon_1' \end{bmatrix},$$

$$\Lambda_3 = \left(\frac{1}{4}\right) \begin{bmatrix} \epsilon_1' \otimes \epsilon_2' \otimes \epsilon_2' \otimes \epsilon_2' \\ \epsilon_2' \otimes \epsilon_1' \otimes \epsilon_2' \otimes \epsilon_2' \\ \epsilon_2' \otimes \epsilon_2' \otimes \epsilon_1' \otimes \epsilon_2' \\ \epsilon_2' \otimes \epsilon_2' \otimes \epsilon_2' \otimes \epsilon_1' \end{bmatrix}.$$

Here $m_i = 2$ for each i. In consideration of Remark 2.3.1, we can work with any choice of $\{P_1, P_2, P_3, P_4\}$, satisfying (2.2.2), for our purpose. Let $P_i = (\frac{1}{\sqrt{2}}, -\frac{1}{\sqrt{2}})$, $1 \le i \le 4$.

If $f = t = 2$, then by (2.3.9), (2.3.17), $P^{(1)} = (\Lambda_1', \Lambda_2')'$ and $\mathcal{I}_d = P^{(1)} R_d P^{(1)'}$ can be seen to be p.d. On the other hand, if $f = 1, t = 3$, then by (2.3.9), (2.3.10), $P^{(1)} = \Lambda_1$, $P^{(2)} = (\Lambda_2', \Lambda_3')'$ and $P^{(2)} R_d P^{(2)'}$ is seen to be invertible so that, by (2.3.17), $\mathcal{I}_d$ equals

$$\left(\frac{1}{14}\right) \begin{bmatrix} 9 & 3 & 3 & 3 & 3 \\ 3 & 1 & 1 & 1 & 1 \\ 3 & 1 & 1 & 1 & 1 \\ 3 & 1 & 1 & 1 & 1 \\ 3 & 1 & 1 & 1 & 1 \end{bmatrix},$$

which is singular. Hence d is a Resolution $(2, 2)$ plan but *not* a Resolution $(1, 3)$ plan. ■

2.5 OPTIMALITY CRITERIA

We continue with the setup of Section 2.3 where under the model (2.3.12) resulting from the absence of factorial effects involving $t+1$ or more factors, one is interested in estimating $\boldsymbol{\beta}^{(1)}$, representing the general mean and complete sets of orthonormal contrasts belonging to factorial effects involving f or less factors ($1 \leq f \leq t \leq n-1$) using an N-run fractional factorial plan. In most practical applications, N is fixed *a priori* from cost considerations. For given $N (0 < N < v)$, let $\mathcal{D}_N$ be the class of all N-run plans. In this and the next section, we consider the issue of finding a fractional factorial plan $d \in \mathcal{D}_N$ which is "good" in some reasonable sense in the present context. The minimal requirement on d is that it must be capable of keeping $\boldsymbol{\beta}^{(1)}$ estimable under (2.3.12), in which case, by Theorem 2.3.1, the information matrix $\mathcal{I}_d$ given by (2.3.17) must be p.d. In other words, we require that d must at least belong to the subclass, ${\mathcal{D}_N}^*$, of $\mathcal{D}_N$, consisting of the N-run Resolution (f, t) plans.

Assume that ${\mathcal{D}_N}^*$ is nonempty. For the case $f = t$, by Theorem 2.3.2, this happens if and only if $N \geq \alpha_f$. How does one compare plans in ${\mathcal{D}_N}^*$? A clue is given by the last part of Theorem 2.3.1 which states that $\mathbb{D}(\hat{\boldsymbol{\beta}}^{(1)}) = \sigma^2 {\mathcal{I}_d}^{-1}$ for each $d \in {\mathcal{D}_N}^*$. We look for a plan that not only ensures estimability of $\boldsymbol{\beta}^{(1)}$ but also leads to an efficient estimation. This calls for choosing $d \in {\mathcal{D}_N}^*$ so as to "minimize" the dispersion of $\hat{\boldsymbol{\beta}}^{(1)}$, as represented by ${\mathcal{I}_d}^{-1}$, in some sensible way. The following commonly used optimality criteria arise naturally. Here, and throughout the book, for a square matrix A, $\det(A)$ denotes its determinant and $\operatorname{tr}(A)$, its trace.

D-Optimality: A D-optimal Resolution (f, t) plan in $\mathcal{D}_N$ is one that belongs to ${\mathcal{D}_N}^*$ and minimizes $\det({\mathcal{I}_d}^{-1})$ over $d \in {\mathcal{D}_N}^*$.

A-Optimality: An A-optimal Resolution (f, t) plan in $\mathcal{D}_N$ is one that belongs to ${\mathcal{D}_N}^*$ and minimizes $\operatorname{tr}({\mathcal{I}_d}^{-1})$ over $d \in {\mathcal{D}_N}^*$.

E-Optimality: An E-optimal Resolution (f, t) plan in $\mathcal{D}_N$ is one that belongs to ${\mathcal{D}_N}^*$ and minimizes the maximum eigenvalue of ${\mathcal{I}_d}^{-1}$ over $d \in {\mathcal{D}_N}^*$.

Thus a D-optimal plan minimizes the generalized variance of $\hat{\boldsymbol{\beta}}^{(1)}$, an A-optimal plan minimizes the average variance of the elements of $\hat{\boldsymbol{\beta}}^{(1)}$, and an E-optimal plan minimizes the maximum possible variance of a normalized linear function of $\hat{\boldsymbol{\beta}}^{(1)}$. If we assume normality of the observational vector $\boldsymbol{Y}$, then an E-optimal plan also maximizes the minimum power of the usual F-test for a null hypothesis of the form $\boldsymbol{\beta}^{(1)} = \boldsymbol{\beta}_0{}^{(1)}$, the minimum being taken along any spherical contour centered at the null hypothetical value. Furthermore, under normality of $\boldsymbol{Y}$, a D-optimal plan minimizes the volume of a confidence ellipsoid for $\boldsymbol{\beta}^{(1)}$ as obtainable by the inversion of such an F-statistic.

The notion of *universal optimality*, due to Kiefer (1975), helps in unifying the various optimality criteria. Since we are concerned with a full rank model represented by a p.d. information matrix $\mathcal{I}_d$ of order α_f, proceeding along the line of Sinha and Mukerjee (1982), the definition of universal optimality can be adapted in our context as follows.

Let $\mathcal{M}$ denote the class of p.d. matrices of order α_f, and for a positive integer ν, let J_ν denote the $\nu \times \nu$ matrix with all elements unity. Consider the class Φ of real-valued functions $\phi(\cdot)$, defined on $\mathcal{M}$, such that

(a) $\phi(\cdot)$ is convex; that is, for every $M_1, M_2 \in \mathcal{M}$, and real a $(0 \le a \le 1)$,

$$a\phi(M_1) + (1-a)\phi(M_2) \ge \phi(aM_1 + (1-a)M_2);$$

(b) $\phi(c_1 I_{\alpha_f} + c_2 J_{\alpha_f}) \ge \phi(c I_{\alpha_f})$ whenever $c \ge c_1 + c_2$, where c_1, c_2 and c are scalars and $c_1 I_{\alpha_f} + c_2 J_{\alpha_f} \in \mathcal{M}$, $c I_{\alpha_f} \in \mathcal{M}$;

(c) $\phi(\cdot)$ is permutation invariant; that is, $\phi(ZMZ') = \phi(M)$, for every $M \in \mathcal{M}$ and every permutation matrix Z of order α_f.

Definition 2.5.1. *A universally optimal Resolution (f, t) plan in $\mathcal{D}_N$ is one that belongs to $\mathcal{D}_N{}^*$ and, for each $\phi(\cdot) \in \Phi$, minimizes $\phi(\mathcal{I}_d)$ over $d \in \mathcal{D}_N{}^*$.*

We refer to Shah and Sinha (1989) and Pukelsheim (1993) for more detailed discussion on various optimality criteria. Consideration of the functions

$$\phi(M) = \log\{\det(M^{-1})\}, \qquad M \in \mathcal{M},$$

$$\phi(M) = \operatorname{tr}(M^{-1}), \qquad M \in \mathcal{M},$$

and

$$\phi(M) = \text{largest eigenvalue of } M^{-1}, \qquad M \in \mathcal{M},$$

which are all members of Φ, shows that a universally optimal plan is also D-, A-, and E-optimal. In the rest of the chapter, we will be considering universally optimal plans based on *orthogonal arrays*, to be introduced in the next section. The following lemmas will be useful in the subsequent development:

Lemma 2.5.1. *For each $d \in \mathcal{D}_N{}^*$, $\operatorname{tr}(\mathcal{I}_d) \le (N/v)\alpha_f$.*

Proof. For any $d \in \mathcal{D}_N{}^*$, by (2.3.17), $P^{(1)}R_d(P^{(1)})' - \mathcal{I}_d$ is nonnegative definite (n.n.d.). Hence by (2.3.9),

$$\mathrm{tr}(\mathcal{I}_d) \le \mathrm{tr}(P^{(1)}R_d(P^{(1)})') = \mathrm{tr}(R_d(P^{(1)})'P^{(1)}) = \sum_{\boldsymbol{x}\in\Omega_f} \mathrm{tr}(R_d(P^{\boldsymbol{x}})'P^{\boldsymbol{x}}). \tag{2.5.1}$$

Now by (2.2.4) and (2.2.5), for each $\boldsymbol{x} \in \Omega(\supset \Omega_f)$,

$$(P^{\boldsymbol{x}})'P^{\boldsymbol{x}} = W^{\boldsymbol{x}} \tag{2.5.2}$$

where

$$W^{\boldsymbol{x}} = \bigotimes_{i=1}^{n} W_i^{x_i}, \tag{2.5.3}$$

and for $1 \le i \le n$,

$$W_i^{x_i} = \begin{cases} m_i^{-1}J_{m_i} & \text{if } x_i = 0 \\ I_{m_i} - m_i^{-1}J_{m_i} & \text{if } x_i = 1. \end{cases} \tag{2.5.4}$$

By (2.5.2)–(2.5.4), for each $\boldsymbol{x} \in \Omega$, the diagonal elements of $(P^{\boldsymbol{x}})'P^{\boldsymbol{x}}$ are all equal to $\prod_{i=1}^{n}\{(m_i - 1)^{x_i}/m_i\}(= v^{-1}\alpha(\boldsymbol{x})$, by (2.2.1)). Hence, recalling that R_d is a diagonal matrix, by (2.3.1), (2.3.11), and (2.5.1), for each $d \in \mathcal{D}_N{}^*$,

$$\mathrm{tr}(\mathcal{I}_d) \le \sum_{\boldsymbol{x}\in\Omega_f} v^{-1}\alpha(\boldsymbol{x})\mathrm{tr}(R_d) = \left(\frac{N}{v}\right)\alpha_f. \qquad \blacksquare$$

Lemma 2.5.2. *Let there exist a plan $d_0(\in \mathcal{D}_N)$ such that $\mathcal{I}_{d_0} = (N/v)I_{\alpha_f}$. Then d_0 is a universally optimal Resolution (f, t) plan in $\mathcal{D}_N$.*

Proof. We use the same arguments as in Kiefer (1975) to prove this result. Clearly, $d_0 \in \mathcal{D}_N{}^*$. Consider any arbitrary $d \in \mathcal{D}_N{}^*$ and any $\phi(\cdot) \in \Phi$. Let $\bar{\mathcal{I}}_d \in \mathcal{M}$ be defined as

$$\bar{\mathcal{I}}_d = \frac{1}{\alpha_f!}\Sigma'\ Z\mathcal{I}_dZ', \tag{2.5.5}$$

where Σ' denotes summation over all the $\alpha_f!$ possible choices of a permutation matrix Z of order α_f. By Properties (a) and (c) of $\phi(\cdot)$, from (2.5.5),

$$\phi(\bar{\mathcal{I}}_d) \le \frac{1}{\alpha_f!}\Sigma'\ \phi(Z\mathcal{I}_dZ') = \phi(\mathcal{I}_d). \tag{2.5.6}$$

By (2.5.5), $\bar{\mathcal{I}}_d$ must have all diagonal elements equal and all off–diagonal elements equal; namely, we must have

$$\bar{\mathcal{I}}_d = c_1 I_{\alpha_f} + c_2 J_{\alpha_f}, \tag{2.5.7}$$

for some scalars c_1, c_2. It follows from (2.5.5) that $\text{tr}(\bar{\mathcal{I}}_d) = \text{tr}(\mathcal{I}_d)$. Hence by (2.5.7) and Lemma 2.5.1, $c_1 + c_2 \leq N/v$. From (2.5.7), property (b) of $\phi(\cdot)$ and the form of $\mathcal{I}_{d_0}$, as specified in the statement of the lemma, one now obtains $\phi(\bar{\mathcal{I}}_d) \geq \phi(\mathcal{I}_{d_0})$. Hence by (2.5.6), $\phi(\mathcal{I}_{d_0}) \leq \phi(\mathcal{I}_d)$, which completes the proof. ■

One can employ Lemma 2.5.2 to check that the 4-run plan presented in Example 2.3.1 is a universally optimal Resolution (1, 1) plan in $\mathcal{D}_4$. A systematic approach in this regard will be developed in the next section.

2.6 ROLE OF ORTHOGONAL ARRAYS

We begin with a definition of orthogonal arrays, due to Rao (1973b).

Definition 2.6.1. *An orthogonal array $OA(N, n, m_1 \times \cdots \times m_n, g)$, having N rows, $n(\geq 2)$ columns, $m_1, \ldots, m_n(\geq 2)$ symbols and strength $g(\leq n)$, is an $N \times n$ array, with elements in the ith column from a set of m_i distinct symbols $(1 \leq i \leq n)$, in which all possible combinations of symbols appear equally often as rows in every $N \times g$ subarray.*

Without loss of generality, the symbols appearing in the ith column of an $OA(N, n, m_1 \times \cdots \times m_n, g)$ may be supposed to be $0, 1, \ldots, m_i - 1$. Note that an orthogonal array of strength g is also of strength $g'(1 \leq g' \leq g)$. If among $m_1, \ldots, m_n$, there are w_i that equal $\mu_i(1 \leq i \leq u)$, where $w_1, \ldots, w_u, \mu_1, \ldots, \mu_u$ are positive integers ($\mu_1, \ldots, \mu_u \geq 2; w_1 + \cdots + w_u = n$), then we will often write $OA(N, n, \mu_1^{w_1} \times \cdots \times \mu_u^{w_u}, g)$ for $OA(N, n, m_1 \times \cdots \times m_n, g)$. In particular, if $m_1 = \cdots = m_n = m$, say, then we get a symmetric orthogonal array (Rao, 1947) which will be denoted simply by $OA(N, n, m, g)$ following the standard notation for such arrays (e.g., see Raghavarao, 1971).

Example 2.6.1. In (2.6.1), (2.6.2), (2.6.3) below, we show an $OA(8, 5, 4 \times 2^4, 2)$, a symmetric $OA(9, 4, 3, 2)$, and a symmetric $OA(8, 4, 2, 3)$.

$$OA(8, 5, 4 \times 2^4, 2) = \begin{bmatrix} 0 & 0 & 0 & 0 & 0 \\ 0 & 1 & 1 & 1 & 1 \\ 1 & 0 & 0 & 1 & 1 \\ 1 & 1 & 1 & 0 & 0 \\ 2 & 0 & 1 & 0 & 1 \\ 2 & 1 & 0 & 1 & 0 \\ 3 & 0 & 1 & 1 & 0 \\ 3 & 1 & 0 & 0 & 1 \end{bmatrix}. \tag{2.6.1}$$

$$OA(9, 4, 3, 2) = \begin{bmatrix} 0 & 0 & 0 & 0 \\ 0 & 1 & 1 & 1 \\ 0 & 2 & 2 & 2 \\ 1 & 0 & 1 & 2 \\ 1 & 1 & 2 & 0 \\ 1 & 2 & 0 & 1 \\ 2 & 0 & 2 & 1 \\ 2 & 1 & 0 & 2 \\ 2 & 2 & 1 & 0 \end{bmatrix}. \tag{2.6.2}$$

$$OA(8, 4, 2, 3) = \begin{bmatrix} 0 & 0 & 0 & 0 \\ 0 & 0 & 1 & 1 \\ 0 & 1 & 0 & 1 \\ 0 & 1 & 1 & 0 \\ 1 & 0 & 0 & 1 \\ 1 & 0 & 1 & 0 \\ 1 & 1 & 0 & 0 \\ 1 & 1 & 1 & 1 \end{bmatrix}. \tag{2.6.3}$$

■

The rows of an $OA(N, n, m_1 \times \cdots \times m_n, g)$ can be identified with the treatment combinations of an $m_1 \times \cdots \times m_n$ factorial setup. Thus the array itself represents an N-run plan for such a factorial. For example, the array $OA(9, 4, 3, 2)$ shown in Equation (2.6.2) represents a 9-run plan for a 3^4 factorial. This is given by

$$\{0000, 0111, 0222, 1012, 1120, 1201, 2021, 2102, 2210\},$$

as obtained by listing the rows of the array. In an $OA(N, n, m_1 \times \cdots \times m_n, g)$, if $g = n$ then, by the definition of an orthogonal array, N must be an integral multiple of $\prod_{i=1}^{n} m_i (= v)$ so that $N \geq v$ and the array cannot represent a fractional plan. For this reason, here we will be primarily concerned with orthogonal arrays for which $g < n$. All the arrays shown in Example 2.6.1 are of this type.

The following lemma plays a crucial role in the rest of this section. Here for each $\boldsymbol{x} = x_1 \ldots x_n \in \Omega$ and $\boldsymbol{y} = y_1 \ldots y_n \in \Omega$, we define $\xi(\boldsymbol{x}) = \prod_{i=1}^{n} m_i^{x_i}$ and $\boldsymbol{x} \vee \boldsymbol{y} = z_1 \ldots z_n$, with $z_i = \max(x_i, y_i)$, $1 \le i \le n$. For example, if $n = 4$, $\boldsymbol{x} = 1100$, $\boldsymbol{y} = 1010$ then $\boldsymbol{x} \vee \boldsymbol{y} = 1110$. Clearly, if $\boldsymbol{x}, \boldsymbol{y} \in \Omega$, then $\boldsymbol{x} \vee \boldsymbol{y} \in \Omega$.

Lemma 2.6.1. *Let $d_0 (\in \mathcal{D}_N)$ be represented by an $OA(N, n, m_1 \times \cdots \times m_n, g)$. Then,*

(a) for each $\boldsymbol{x} \in \Omega_g$, $P^{\boldsymbol{x}} R_{d_0} (P^{\boldsymbol{x}})' = (N/v) I_{\alpha(\boldsymbol{x})}$,

(b) for each $\boldsymbol{x}, \boldsymbol{y} \in \Omega_g$ such that $\boldsymbol{x} \vee \boldsymbol{y} \in \Omega_g$ and $\boldsymbol{x} \neq \boldsymbol{y}$, $P^{\boldsymbol{x}} R_{d_0} (P^{\boldsymbol{y}})' = O$.

Proof. Consider any $\boldsymbol{x}, \boldsymbol{y} \in \Omega_g$ such that $\boldsymbol{x} \vee \boldsymbol{y} = \boldsymbol{z} = z_1 z_2 \ldots z_n \in \Omega_g$. At this stage nothing is assumed regarding the equality or otherwise of $\boldsymbol{x}$ and $\boldsymbol{y}$. Let S denote the set of subscripts i such that $z_i = 1$ and $\bar{S}$ be the complement of S in $\{1, \ldots, n\}$. Define the $\xi(\boldsymbol{z}) \times v$ matrix

$$U^{\boldsymbol{z}} = \bigotimes_{i=1}^{n} U_i^{z_i}, \tag{2.6.4}$$

where for $1 \le i \le n$,

$$U_i^{z_i} = \begin{cases} \mathbf{1}'_{m_i} & \text{if } z_i = 0 \\ I_{m_i} & \text{if } z_i = 1. \end{cases} \tag{2.6.5}$$

By (2.6.4), (2.6.5), and the definition of R_{d_0}, $U^{\boldsymbol{z}} R_{d_0} (U^{\boldsymbol{z}})'$ is a diagonal matrix, each diagonal element being given by a partial sum of the quantities $r_{d_0}(j_1 \ldots j_n)$, where summation extends over $j_i, i \in \bar{S}$ with $j_i, i \in S$ held fixed. Now, $\boldsymbol{z} \in \Omega_g$ (i.e., S has cardinality at most g) and d_0 is represented by an $OA(N, n, m_1 \times \cdots \times m_n, g)$. Hence, recalling the definition of an orthogonal array, it follows that

$$U^{\boldsymbol{z}} R_{d_0} (U^{\boldsymbol{z}})' = \left\{ \frac{N}{\xi(\boldsymbol{z})} \right\} I_{\xi(\boldsymbol{z})}. \tag{2.6.6}$$

By the definition of $\boldsymbol{z}$, if $z_i = 0$, then $x_i = y_i = 0$ $(1 \le i \le n)$. Hence from (2.2.4), (2.2.5), (2.6.4) and (2.6.5),

$$P^{\boldsymbol{x}} = \left\{ \frac{\xi(\boldsymbol{x})}{v} \right\}^{1/2} Q(\boldsymbol{x}, \boldsymbol{z}) U^{\boldsymbol{z}}, \quad P^{\boldsymbol{y}} = \left\{ \frac{\xi(\boldsymbol{y})}{v} \right\}^{1/2} Q(\boldsymbol{y}, \boldsymbol{z}) U^{\boldsymbol{z}}, \tag{2.6.7}$$

where

$$Q(\boldsymbol{x}, \boldsymbol{z}) = \bigotimes_{i=1}^{n} Q_i(x_i, z_i), \quad Q(\boldsymbol{y}, \boldsymbol{z}) = \bigotimes_{i=1}^{n} Q_i(y_i, z_i), \tag{2.6.8}$$

with

$$Q_i(0,0) = 1, \quad Q_i(0,1) = \mathbf{1}'_{m_i}, \quad Q_i(1,1) = P_i, \quad 1 \le i \le n. \tag{2.6.9}$$

By (2.6.6)–(2.6.8), if $\boldsymbol{x}, \boldsymbol{y} (\in \Omega_g)$ are such that $\boldsymbol{z} = \boldsymbol{x} \vee \boldsymbol{y} \in \Omega_g$, then

$$P^{\boldsymbol{x}} R_{d_0} (P^{\boldsymbol{y}})' = \left[\frac{N\{\xi(\boldsymbol{x})\xi(\boldsymbol{y})\}^{1/2}}{v\xi(\boldsymbol{z})} \right] \bigotimes_{i=1}^{n} Q_i(x_i, z_i) Q_i(y_i, z_i)'. \tag{2.6.10}$$

Now, if $\boldsymbol{x} = x_1 \ldots x_n$ is different from $\boldsymbol{y} = y_1 \ldots y_n$, then for some i, we must have either $x_i = 0$, $y_i = 1$, $z_i = 1$ or $x_i = 1$, $y_i = 0$, and $z_i = 1$. By (2.2.2) and (2.6.9), for this i, $Q_i(x_i, z_i) Q_i(y_i, z_i)'$ equals the null row or column vector, which proves (b) in consideration of (2.6.10). Similarly, for $\boldsymbol{x} \in \Omega_g$, taking $\boldsymbol{y} = \boldsymbol{x}$ in (2.6.10) so that $\boldsymbol{z} = \boldsymbol{x}$, part (a) of the lemma follows easily from (2.2.2) and (2.6.9). ∎

We now present the main result of this section, which shows the universal optimality of plans represented by orthogonal arrays. This was first proved by Cheng (1980b) for the case $g = 2$ and later extended by Mukerjee (1982) to the case of general g. It covers many of the earlier findings in Kounias (1977) who worked with 2^n factorials and specific optimality criteria.

Theorem 2.6.1. *Let $d_0 (\in \mathcal{D}_N)$ be represented by an $OA(N, n, m_1 \times \cdots \times m_n, g)$, where $2 \le g \le n$. Then d_0 is a universally optimal Resolution (f, t) plan in $\mathcal{D}_N$ for every choice of integers f, t such that $f + t = g$ and $1 \le f \le t \le n - 1$.*

Proof. Let f and t be as in the statement of the theorem. Then $\Omega_f \subset \Omega_t \subset \Omega_g$ and $\boldsymbol{x} \vee \boldsymbol{y} \in \Omega_g$ for every $\boldsymbol{x} \in \Omega_f$ and $\boldsymbol{y} \in \Omega_t$. Hence by Lemma 2.6.1,

$$P^{\boldsymbol{x}} R_{d_0} (P^{\boldsymbol{x}})' = \left(\frac{N}{v}\right) I_{\alpha(\boldsymbol{x})}, \qquad \text{for each } \boldsymbol{x} \in \Omega_f,$$

$$P^{\boldsymbol{x}} R_{d_0} (P^{\boldsymbol{y}})' = O \qquad \text{for each } \boldsymbol{x} \in \Omega_f, \boldsymbol{y} \in \Omega_t, \boldsymbol{x} \neq \boldsymbol{y}.$$

Recalling the definitions of $P^{(1)}$ and $P^{(2)}$ (*vide* (2.3.9) and (2.3.10)), one now obtains

$$P^{(1)} R_{d_0} (P^{(1)})' = \left(\frac{N}{v}\right) I_{\alpha_f}, \; P^{(1)} R_{d_0} (P^{(2)})' = O, \tag{2.6.11}$$

so that by (2.3.17),

$$\mathcal{I}_{d_0} = \left(\frac{N}{v}\right) I_{\alpha_f}, \tag{2.6.12}$$

and the result follows from Lemma 2.5.2. ■

By Theorem 2.6.1, the orthogonal arrays displayed in (2.6.1), (2.6.2), and (2.6.3) represent (a) a universally optimal Resolution (1, 1) plan in $\mathcal{D}_8$ for a 4×2^4 factorial, (b) a universally optimal Resolution (1, 1) plan in $\mathcal{D}_9$ for a 3^4 factorial and (c) a universally optimal Resolution (1, 2) plan in $\mathcal{D}_8$ for a 2^4 factorial. Also, as one can easily check, the 4-run plan exhibited in Example 2.3.1 is represented by a symmetric $OA(4, 3, 2, 2)$. Therefore, as noted in the last section, it is a universally optimal Resolution (1, 1) plan in $\mathcal{D}_4$ for a 2^3 factorial.

Remark 2.6.1.

(a) Let $d_0 \in \mathcal{D}_N$ be represented by an orthogonal array as in Theorem 2.6.1. Then for any integers f, t satisfying $f + t = g$ and $1 \leq f \leq t \leq n - 1$, by (2.6.12) and Theorem 2.3.1,

$$\mathbb{D}(\hat{\boldsymbol{\beta}}^{(1)}) = \sigma^2 \left(\frac{v}{N}\right) I_{\alpha_f}; \tag{2.6.13}$$

namely, the elements of $\hat{\boldsymbol{\beta}}^{(1)}$ are uncorrelated.

(b) Furthermore, proceeding as in the proof of Theorem 2.3.1 or even otherwise, it is not difficult to see from (2.6.11) that

$$\hat{\boldsymbol{\beta}}^{(1)} = \left(\frac{v}{N}\right) P^{(1)} X'_{d_0} \boldsymbol{Y}, \tag{2.6.14}$$

where X_{d_0} is the design matrix for d_0 (see (2.3.2)) and $\boldsymbol{Y}$ is the observational vector. Consequently the BLUE of every element of $\boldsymbol{\beta}^{(1)}$ is the same as what would be obtained if $\boldsymbol{\beta}^{(2)}$ and the other elements of $\boldsymbol{\beta}^{(1)}$ were not included in the model (2.3.12). In other words, d_0 allows estimation of each element of $\boldsymbol{\beta}^{(1)}$ in a manner which is not affected by $\boldsymbol{\beta}^{(2)}$ and other elements of $\boldsymbol{\beta}^{(1)}$.

(c) The relations (2.6.13) and (2.6.14) greatly simplify the analysis of the data arising from d_0. ■

Remark 2.6.2. In consideration of Remark 2.6.1(a) and (b), following Box and Hunter (1961a, b), a plan represented by an orthogonal array of strength

g is called an orthogonal Resolution $(g + 1)$ plan. The use of the phrase "Resolution $(g + 1)$" is now meaningful since by Theorem 2.6.1, such a plan is of Resolution (f, t), whenever $f + t = g$ and $1 \le f \le t \le n - 1$. This may be contrasted with what was discussed earlier in Section 2.4 in a more general context. Since an orthogonal array of strength g is also of strength $g'(\le g)$, it follows that an orthogonal Resolution $(g + 1)$ plan is also an orthogonal Resolution $(g' + 1)$ plan, where $g' \le g$. ■

Apart from being useful in the proof of Theorem 2.6.1, Lemma 2.6.1 leads to Rao's bound, a very important necessary condition for the existence of an orthogonal array. For $1 \le i \le n$, $1 \le \nu \le n - 1$, let $\Omega_{i\nu}$ be a subset of Ω consisting of those binary n-tuples $x_1 \ldots x_n$ for which x_i and also exactly ν of $x_1, \ldots, x_{i-1}, x_{i+1}, \ldots, x_n$ equal unity. Thus, if $n = 4$, then

$$\Omega_{12} = \{1110, 1101, 1011\}$$

and

$$\Omega_{21} = \{1100, 0110, 0101\}.$$

Let for $1 \le i \le n$, $1 \le \nu \le n - 1$,

$$\alpha_{i\nu} = \sum_{x \in \Omega_{i\nu}} \alpha(x), \qquad \alpha_\nu{}^* = \max\{\alpha_{1\nu}, \ldots, \alpha_{n\nu}\}. \tag{2.6.15}$$

Also let α_ν be as in (2.3.11).

Theorem 2.6.2. *(Rao's bound) Suppose that an orthogonal array $OA(N, n, m_1 \times \cdots \times m_n, g)$ exists, where $2 \le g \le n$.*

(a) If $g(= 2s, s \ge 1)$ is even, then $N \ge \alpha_s$.
(b) If $g(= 2s + 1, s \ge 1)$ is odd, then $N \ge \alpha_s + \alpha_s{}^*$.

Proof.

(a) In this case, by Theorem 2.6.1, the orthogonal array under consideration represents an N-run Resolution (s, s) plan for an $m_1 \times \cdots \times m_n$ factorial. Hence by Theorem 2.3.2, $N \ge \alpha_s$.

(b) In this case, for $1 \le i \le n$, define the matrix B_i, with $\alpha_s + \alpha_{is}$ rows and v columns, as

$$B_i = (\ldots, (P^x)', \ldots)'_{x \in \Omega_s \cup \Omega_{is}}. \tag{2.6.16}$$

For instance, if $n = 4$ and $g = 3$, then $s = 1$ and

$$B_2 = (P^{0000'}, P^{1000'}, P^{0100'}, P^{0010'}, P^{0001'}, P^{1100'}, P^{0110'}, P^{0101'})'.$$

Observe that $\Omega_s \cup \Omega_{is} \subset \Omega_g$ and that if $\boldsymbol{x}, \boldsymbol{y} \in \Omega_s \cup \Omega_{is}$ then $\boldsymbol{x} \vee \boldsymbol{y} \in \Omega_g$. Hence if $d_0(\in \mathcal{D}_N)$ be represented by the orthogonal array under consideration then by (2.6.16) and Lemma 2.6.1, $B_i R_{d_0} B_i' = (N/v)I$ where I is an identity matrix of order $\alpha_s + \alpha_{is}$. Thus Rank$(R_{d_0}) \geq \alpha_s + \alpha_{is}$ and as in the proof of Theorem 2.3.2, $N \geq \alpha_s + \alpha_{is}$ $(1 \leq i \leq n)$. Consequently, by (2.6.15), $N \geq \alpha_s + \alpha_s{}^*$. ■

Remark 2.6.3. In particular, if $m_1 = \cdots = m_n = m$ then by (2.2.1), $\alpha(\boldsymbol{x}) = (m-1)^u$ for any $\boldsymbol{x}(\in \Omega)$ with exactly u components equal to unity. Hence by (2.3.11), (2.6.15) and Theorem 2.6.2, for a symmetric $OA(N, n, m, g)$ we have

(a) $N \geq \sum_{u=0}^{s} \binom{n}{u}(m-1)^u$ if $g(= 2s, s \geq 1)$ is even,

(b) $N \geq \sum_{u=0}^{s} \binom{n}{u}(m-1)^u + \binom{n-1}{s}(m-1)^{s+1}$ if $g(= 2s+1, s \geq 1)$ is odd.

These give Rao's (1947) bound as obtained originally for symmetric orthogonal arrays. ■

Remark 2.6.4. Theorem 2.6.2 sets a lower bound on the number of runs in a plan represented by an orthogonal array. An orthogonal array attaining Rao's bound, as shown in this theorem, is known as *complete* or *tight*. For even g, such an array is also called saturated in keeping with the notion of a saturated fractional factorial plan. The orthogonal arrays shown in Example 2.6.1 are all tight. It may be noted that Rao's bound provides only a necessary but not sufficient condition for the existence of an orthogonal array. In particular, for a specific set of values of $n, m_1, \ldots, m_n, g$, a tight orthogonal array may or may not exist. We will return to this point in Chapter 5. ■

In view of Remark 2.6.1 and, more important, Theorem 2.6.1, orthogonal arrays have been very popular as fractional factorial plans. We therefore find it appropriate to discuss in some detail the issues relating to the construction and existence of orthogonal arrays. This will be done in Chapters 3, 4, and 5.

EXERCISES

2.1. Consider the setup of a $2 \times 3 \times 4$ factorial experiment.

(a) Write down a specific choice of the matrices P_1, P_2 and P_3, as defined in (2.2.2).

(b) Hence use (2.2.4) to find explicit expressions for the matrices $P^{\boldsymbol{x}}$, $\boldsymbol{x} \in \Omega$.

2.2. In a $2 \times 3 \times 4$ factorial experiment, let all interactions be absent. Suppose that interest lies only in the general mean and contrasts belonging to the main effects.

(a) Use the result of Exercise 2.1 to obtain an explicit expression for the matrix $P^{(1)}$, as defined in (2.3.9).

(b) Hence proceed as in the proof of Theorem 2.3.2 to obtain a Resolution (1, 1) plan in seven runs.

(c) In the setup under consideration, is it possible to find a Resolution (1, 1) plan in six or less runs?

2.3. Verify the statement made in the concluding sentence of the paragraph just below the proof of Theorem 2.3.2.

2.4. With reference to an $m_1 \times \cdots \times m_n$ factorial, suppose interest lies in the general mean and contrasts belonging to the main effects under the absence of all interactions. Let d be a plan consisting of $N = 1 + \sum_{i=1}^{n}(m_i - 1)$ runs as shown below:

$$\begin{array}{ll} 00\ldots0, & \\ i_1 0\ldots0, & 1 \le i_1 \le m_1 - 1, \\ 0 i_2 \ldots 0, & 1 \le i_2 \le m_2 - 1, \\ \vdots & \\ 00\ldots0 i_n, & 1 \le i_n \le m_n - 1. \end{array}$$

For example, if $n = 2$, $m_1 = 3$ and $m_2 = 2$, then $d = \{00, 10, 20, 01\}$.

(a) Taking $f = t = 1$ in the setup of Section 2.3, use (2.3.9) and (2.3.13) to express the matrix X_{1d} in terms of the columns of the matrices $P_i, \ldots, P_n$ introduced in (2.2.2).

(b) Show that the columns of X_{1d} are linearly independent and hence conclude that d is a Resolution (1, 1) plan.

(c) As a Resolution (1, 1) plan, is d necessarily D-, A-, or E-optimal within the class of plans involving the same number of runs? [*Hint*:

Take $n = 3$, $m_1 = m_2 = m_3 = 2$ and compare d with the plan shown in Example 2.3.1.]

2.5. Example 2.4.1 illustrates a Resolution (2, 2) plan which is not a Resolution (1, 3) plan. Construct another such example.

2.6. Under the setup of Section 2.6, verify that the BLUE of $\boldsymbol{\beta}^{(1)}$ is given by (2.6.14).

CHAPTER 3

Symmetric Orthogonal Arrays

3.1 INTRODUCTION

In Chapter 2 we introduced orthogonal arrays and discussed their utility and importance in connection with fractional factorial plans. This chapter deals with the construction of *symmetric* orthogonal arrays. For ease in reference, recall from Section 2.6 that a symmetric orthogonal array, $OA(N, n, m, g)$ is an $N \times n$ array, with elements from a set of $m(\geq 2)$ distinct symbols, in which all possible combinations of symbols appear equally often as rows in every $N \times g(g \leq n)$ subarray.

We begin the discussion in Section 3.2 by linking two-symbol orthogonal arrays of strength two with *Hadamard matrices*. Section 3.3 presents the "foldover" technique of obtaining two-symbol orthogonal arrays of *odd* strength from those of even strength. Sections 3.4 and 3.5 deal, respectively, with the use of Galois fields and difference matrices for the construction of symmetric orthogonal arrays in general. Finally some miscellaneous results are briefly discussed in Section 3.6.

Deletion of any column in an orthogonal array $OA(N, n, m, g)$, with $g < n$, yields an $OA(N, n-1, m, g)$. Hence, as far as practicable, we attempt to present procedures of construction which lead to arrays with the maximum possible value of n for given values of the other parameters. Clearly this objective is attained, in particular, for procedures that yield tight orthogonal arrays (see Remarks 2.6.3 and 2.6.4).

3.2 ORTHOGONAL ARRAYS AND HADAMARD MATRICES

To begin with, we have the following definition:

Definition 3.2.1. *A square matrix H_N of order N and entries ± 1 is called a Hadamard matrix if $H_N' H_N = N I_N$, where I_N is the Nth order identity matrix.*

If H_N is a Hadamard matrix of order N, then it follows from the above definition that $H_N H_N' = N I_N$. Also it is easy to see that if any row or column of a Hadamard matrix is multiplied by -1, the resultant matrix remains Hadamard. Thus without loss of generality, one can write a Hadamard matrix with its first row and first column consisting of only +1's. A Hadamard matrix with its first row and first column having only +1's is said to be in its *normal* form. Also we say that a Hadamard matrix is in its *seminormal* form if its first column contains only +1's.

Trivially H_N exists when $N = 1$ and H_2 is given by

$$H_2 = \begin{pmatrix} 1 & 1 \\ 1 & -1 \end{pmatrix}. \tag{3.2.1}$$

We now present a result that helps in studying the existence of H_N for other values of N.

Theorem 3.2.1. *A necessary condition for the existence of a Hadamard matrix of order $N > 2$ is that $N \equiv 0(mod\ 4)$.*

Proof. Suppose that H_N exists for some $N > 2$. Without loss of generality, we can write the first row of H_N with only +1's. Consider the $3 \times N$ submatrix of H_N given by its first three rows, displayed as

$$\left(\begin{array}{ccc|ccc|ccc|ccc} 1\ 1 & \dots & 1 & 1\ \ 1 & \dots & 1 & 1\ \ 1 & \dots & 1 & 1\ \ 1 & \dots & 1 \\ 1\ 1 & \dots & 1 & 1\ \ 1 & \dots & 1 & -1\ -1 & \dots & -1 & -1\ -1 & \dots & -1 \\ 1\ 1 & \dots & 1 & -1\ -1 & \dots & -1 & 1\ \ 1 & \dots & 1 & -1\ -1 & \dots & -1 \end{array}\right).$$

Let x_1, x_2, x_3, x_4 denote, respectively, the frequencies with which the column vectors $(1, 1, 1)'$, $(1, 1, -1)'$, $(1, -1, 1)'$, and $(1, -1, -1)'$ appear as columns in the above submatrix. Then, by the properties of a Hadamard matrix, one obtains

$$\begin{aligned} x_1 + x_2 + x_3 + x_4 &= N, \\ x_1 + x_2 - x_3 - x_4 &= 0, \\ x_1 - x_2 + x_3 - x_4 &= 0, \\ x_1 - x_2 - x_3 + x_4 &= 0. \end{aligned}$$

The (unique) solution to the above equations is $x_1 = x_2 = x_3 = x_4 = N/4$. ■

Theorem 3.2.1 gives only a necessary condition for the existence of a Hadamard matrix. It is not known whether the condition $N \equiv 0(\text{mod } 4)$ is also sufficient for the existence of a Hadamard matrix of order $N > 2$. Although it is widely believed that this condition is necessary and sufficient, no proof of the sufficiency is as yet available.

Hadamard matrices for all permissible values of $N \leq 100$, with the exception of $N = 92$ are displayed in Plackett and Burman (1946). A Hadamard matrix of order 92 was discovered by Baumert, Golomb, and Hall (1962). Hadamard matrices for all permissible values of $N \leq 424$ are now known to exist (e.g., see Hedayat and Wallis, 1978, and Sawade, 1985). For ready reference, the construction of all Hadamard matrices of orders 50 or less has been indicated in the appendix.

The next lemma follows easily from the properties of Kronecker product of matrices (*vide* Section 2.2).

Lemma 3.2.1. *If H_{N_1} and H_{N_2} are Hadamard matrices of orders N_1 and N_2, respectively, then $H_{N_1} \otimes H_{N_2}$ is a Hadamard matrix of order $N_1 N_2$.* ■

Thus a Hadamard matrix of order $2^r, r \geq 2$ always exists, obtained by taking the Kronecker product of H_2 with itself, r times. There are several other methods of construction of Hadamard matrices. For the details, we refer the interested reader to Hall (1986, ch. 14) and to an excellent review paper of Hedayat and Wallis (1978).

A Hadamard matrix of order $N(\geq 4)$ is equivalent to a (symmetric) two-symbol tight orthogonal array of strength two, N rows and $N - 1$ columns. Suppose that H_N is a Hadamard matrix of order N and, without loss of generality, let H_N be written in its seminormal form. Then, by deleting the first column of all +1's from H_N, we get an $N \times (N - 1)$ matrix, with entries +1 and -1. It is easy to see that this matrix represents an orthogonal array $OA(N, N-1, 2, 2)$. Conversely, if there exists an $OA(N, N-1, 2, 2)$, then replacing the two symbols in the array (if necessary) by $+1$ and -1, respectively, and augmenting the resultant $N \times (N - 1)$ matrix by a column of all +1's, one gets a Hadamard matrix of order N. Hence we have the following result:

Theorem 3.2.2. *The existence of a Hadamard matrix of order $N(\geq 4)$ is equivalent to the existence of an orthogonal array $OA(N, N-1, 2, 2)$.* ■

3.3 FOLDOVER TECHNIQUE

Two-symbol orthogonal arrays of *odd* strength can be constructed from those of *even* strength by following a procedure called *foldover technique*. This technique, summarized in Theorem 3.3.1 below, is originally due to Box and Wilson (1951)—see also Hedayat and Stufken (1988).

Theorem 3.3.1. *For positive integral s, the existence of an $OA(N, n, 2, 2s)$ is equivalent to that of an $OA(2N, n+1, 2, 2s+1)$.*

Proof. First suppose that an $OA(N, n, 2, 2s)$ exists. Denote this array by A, and without loss of generality, let its symbols be 0 and 1. Further let $\bar{A}$ be the array obtained by interchanging the symbols 0 and 1 in A. Consider the array

$$A_1 = \begin{bmatrix} A & \mathbf{0} \\ \bar{A} & \mathbf{1} \end{bmatrix}, \tag{3.3.1}$$

where $\mathbf{0}$ and $\mathbf{1}$, respectively, denote vectors of all zeros and all ones, both of order $N \times 1$. We will show that A_1 is an orthogonal array $OA(2N, n+1, 2, 2s+1)$.

To that effect, consider any $N \times (2s+1)$ subarray, say A^*, of A. Let $\bar{A}^*$ be the corresponding $N \times (2s+1)$ subarray of $\bar{A}$. Since A is an $OA(N, n, 2, 2s)$, there are $N/2^{2s}$ rows of A^* which equal either $00\ldots0$ or $10\ldots0$. Hence, if $00\ldots0$ appears x times as a row of A^*, then $10\ldots0$ appears $N/2^{2s} - x$ times as a row of A^*. In general, then, a similar argument shows that every $(2s+1)$-tuple appears as a row of A^* with frequency x if it contains an odd number of zeros and with frequency $N/2^{2s} - x$, if it contains an even number of zeros. Therefore each possible $(2s+1)$-tuple appears $N/2^{2s}$ times as a row of $[A^{*\prime}\bar{A}^{*\prime}]'$. This shows that the array formed by the first n columns of A_1 is an $OA(2N, n, 2, 2s+1)$. Since both A and $\bar{A}$ are orthogonal arrays, $OA(N, n, 2, 2s)$, it now follows that A_1 is an $OA(2N, n+1, 2, 2s+1)$.

Conversely, if an $OA(2N, n+1, 2, 2s+1)$, say B, exists then denoting its symbols by 0 and 1 and permuting its rows (if necessary), B can be expressed as

$$B = \begin{bmatrix} B_1 & \mathbf{0} \\ B_2 & \mathbf{1} \end{bmatrix},$$

where $\mathbf{0}$ and $\mathbf{1}$ are as in the context of (3.3.1). Then it is easy to see that B_1 is an $OA(N, n, 2, 2s)$. ■

The nomenclature "foldover" is derived from the fact that in the construction given by (3.3.1), the first N runs are folded over to get the next N runs. From Theorems 3.2.2 and 3.3.1, the following result is evident:

Corollary 3.3.1. *For $N \geq 4$, the existence of a Hadamard matrix of order N implies the existence of an orthogonal array $OA(2N, N, 2, 3)$.* ■

Remark 3.3.1. Along the lines of the arguments leading to Theorem 3.3.1, it is not hard to see that given H_N, a Hadamard matrix of order $N(\geq 4)$, an $OA(2N, N, 2, 3)$ with symbols $+1$ and -1 can be constructed simply as $[H'_N, -H'_N]'$. In particular, $[H'_4, -H'_4]'$, where $H_4 = H_2 \otimes H_2$ with H_2 given by (3.2.1), represents an $OA(8, 4, 2, 3)$ which is isomorphic to the one shown in (2.6.3). (Two orthogonal arrays are called isomorphic to each other if any one of them can be obtained from the other by permuting the rows and columns and renaming the symbols.) ■

Saha (1975) and Chakravarty and Dey (1976) discussed the connection between certain balanced incomplete block designs and two-symbol orthogonal arrays of strength three. Orthogonal arrays with the same parameters, over a practically meaningful range, however, can be obtained from Corollary 3.3.1 as well.

3.4 USE OF GALOIS FIELDS

We now consider the use of Galois (or, finite) fields for the construction of symmetric orthogonal arrays; a reader unfamiliar with Galois fields may refer, for example, to Raghavarao (1971, pp. 338–341).

Let $m(\geq 2)$ be a prime or a prime power, and consider $GF(m)$, the Galois field of order m. The following fundamental lemma, due to Bose and Bush (1952), will be helpful throughout this section:

Lemma 3.4.1. *Let there exist an $r \times n$ matrix C, with elements from $GF(m)$, such that every $r \times g$ submatrix of C has rank g. Then an orthogonal array $OA(m^r, n, m, g)$ can be constructed.*

Proof. Let $\boldsymbol{\xi}$ denote a typical $r \times 1$ vector with entries from $GF(m)$. Considering all the m^r possible distinct choices of $\boldsymbol{\xi}$, form an $m^r \times n$ array A with rows of the form $\boldsymbol{\xi}'C$. We claim that A is an orthogonal array, $OA(m^r, n, m, g)$ with elements given by the members of $GF(m)$. To see this, consider any $m^r \times g$ subarray, say A_1, of A, and let C_1 be the corresponding $r \times g$ submatrix of C. The rows of A_1 are of the form $\boldsymbol{\xi}'C_1$, and since, by assumption, C_1 has full column rank, there are m^{r-g} distinct choices of $\boldsymbol{\xi}$ such that $\boldsymbol{\xi}'C_1$ equals any fixed g-component row vector with elements from $GF(m)$. Thus in A_1 each possible g-component row vector appears with frequency m^{r-g}, and our claim is established. ■

The next few results, based on specific choices of C, indicate applications of Lemma 3.4.1. We will denote the elements of $GF(m)$ by $\rho_0, \rho_1, \ldots, \rho_{m-1}$. In particular, ρ_0 and ρ_1 will stand for the identity elements with respect to the operations of addition and multiplication, respectively.

Theorem 3.4.1. *(Bush, 1952)*

(a) *Let $m(\geq 2)$ be a prime or a prime power and $g(2 \leq g \leq m)$ be an integer. Then an orthogonal array $OA(m^g, m+1, m, g)$ can be constructed.*

(b) *If in addition $m(\geq 4)$ is even, then an orthogonal array $OA(m^3, m+2, m, 3)$ can be constructed.*

Proof.

(a) In Lemma 3.4.1, take $r = g$, $n = m + 1$ and choose C as the $g \times (m+1)$ matrix with columns $(\rho_1, \rho_j, {\rho_j}^2, \ldots, {\rho_j}^{g-1})'$, $(0 \leq j \leq m-1)$ and $(\rho_0, \rho_0, \ldots, \rho_0, \rho_1)'$. Since the determinant of every $g \times g$ submatrix of C equals a Vandermonde's determinant (e.g., see Rao, 1973a, p. 32), and since $\rho_0, \rho_1, \ldots, \rho_{m-1}$ are distinct, it follows that every $g \times g$ submatrix of C has rank g. The construction is now possible as in Lemma 3.4.1.

(b) If in addition $m(\geq 4)$ is even, then ${\rho_0}^2, {\rho_1}^2, \ldots, {\rho_{m-1}}^2$ are distinct, and the result follows as in part (a) by taking

$$C = \begin{bmatrix} \rho_1 & \rho_1 & \cdots & \rho_1 & \rho_0 & \rho_0 \\ \rho_0 & \rho_1 & \cdots & \rho_{m-1} & \rho_0 & \rho_1 \\ {\rho_0}^2 & {\rho_1}^2 & \cdots & {\rho_{m-1}}^2 & \rho_1 & \rho_0 \end{bmatrix}. \qquad \blacksquare$$

Theorem 3.4.2. *(Bush, 1952) Let $m(\geq 2)$ be a prime or a prime power and $g(\geq 2)$ be an integer. Then an orthogonal array $OA(m^g, g+1, m, g)$ can be constructed.*

Proof. In Lemma 3.4.1, we choose C as the $g \times (g+1)$ matrix

$$[\text{diag}(\rho_1, \rho_1, \ldots, \rho_1)\ \boldsymbol{\epsilon}],$$

where $\boldsymbol{\epsilon} = (\rho_1, \ldots, \rho_1)'$, and note that every $g \times g$ submatrix of C is nonsingular. ■

Theorem 3.4.3. *(Bose and Bush, 1952) Let $m(\geq 2)$ be a prime or a prime power and $r(\geq 2)$ be an integer. Then an orthogonal array $OA(m^r, (m^r - 1)/(m-1), m, 2)$ can be constructed.*

Proof. For $1 \leq j \leq r$, let T_j be a set consisting of those $r \times 1$ vectors, defined over $GF(m)$, that have the first $j - 1$ elements equal to ρ_0 and the

jth element equal to ρ_1. The sets $T_1, T_2, \ldots, T_r$ are disjoint, and for each j, the cardinality of T_j is m^{r-j}. The union, T, of $T_1, \ldots, T_r$ has cardinality $(m^r - 1)/(m - 1)$, and it is easy to see that every two distinct vectors in T are linearly independent. Hence the result follows from Lemma 3.4.1 taking C as an $r \times \{(m^r - 1)/(m - 1)\}$ matrix with columns given by the vectors in T. ■

Remark 3.4.1. Orthogonal arrays, as obtained from Theorems 3.4.1(b) and 3.4.3 are tight (*vide* Remarks 2.6.3 and 2.6.4). Also it can be shown (see Bush, 1952; Raghavarao, 1971, ch. 2) that

(i) in an $OA(m^g, n, m, g)$, if $m \leq g$, then $n \leq g + 1$, and
(ii) in an $OA(m^3, n, m, 3)$, if m is odd then $n \leq m + 1$.

Theorems 3.4.2 and 3.4.1(a) lead to the attainment of these upper bounds on n when m is a prime or a prime power. ■

Remark 3.4.2. Taking $g = 2$ in Theorem 3.4.1(a) or $r = 2$ in Theorem 3.4.3, one gets an $OA(m^2, m + 1, m, 2)$ which could as well be obtained from Latin squares. We find it appropriate to discuss this point in a little more detail. A Latin square of order $m(\geq 2)$ is an $m \times m$ array, with entries from a set of m distinct symbols such that each symbol appears exactly once in each row and each column of the array. Two Latin squares of the same order are said to be orthogonal to each other if, when one of the squares is superimposed on the other, every ordered pair of elements appears precisely once. A set of Latin squares is said to form a set of mutually orthogonal Latin squares (MOLS) if every pair in the set is orthogonal to each other. The maximum number of MOLS of order $m(\geq 2)$ is $(m - 1)$, this number being attainable if m is a prime or a prime power (e.g., see Raghavarao, 1971, ch. 1).

In other words, if $m(\geq 2)$ is a prime or prime power, then a set of $m - 1$ MOLS of order m is available. Without loss of generality, let the symbols in these Latin squares be $0, 1, \ldots, m - 1$. Superimpose all the Latin squares in the set of MOLS one over the other, so as to have an arrangement of m rows and m columns, each cell in the arrangement containing an ordered $(m - 1)$-tuple of symbols. Label the rows and columns of the arrangement also by the numbers $0, 1, \ldots, m - 1$. From the m^2 cells of this arrangement, generate m^2 rows by augmenting each cell entry by the row and column indices. The $m^2 \times (m + 1)$ array so formed can be easily verified to be an $OA(m^2, m + 1, m, 2)$.

If $m(\geq 2)$ is not a prime or prime power but $k(< m - 1)$ MOLS of order m are available, then the same procedure leads to an orthogonal array $OA(m^2, k + 2, m, 2)$. ■

Remark 3.4.3. Readers familiar with finite projective geometry will recognize the vectors in the set T, used in proving Theorem 3.4.3, as distinct points in $PG(r-1, m)$, the finite $(r-1)$-dimensional projective geometry defined over $GF(m)$ (e.g., see Raghavarao, 1971, pp. 357–359). More generally, in Lemma 3.4.1, the columns of the matrix C satisfying the rank condition can be interpreted as points in $PG(r-1, m)$, such that no g of them are conjoint or linearly dependent. Hence Lemma 3.4.1 means that if n points can be found in $PG(r-1, m)$ such that no g of them are conjoint then an $OA(m^r, n, m, g)$ can be constructed (Bose and Bush, 1952). In general, for arbitrary m, r, n, and g, the problem of finding n such points is quite deep and linked with the so-called *packing problem* in confounded symmetrical factorial experiments. We refer the reader to Bose (1947) for more details on this topic. Because of the connection with finite projective geometry as indicated above, the construction procedure arising from Lemma 3.4.1 is often referred to as a geometric procedure. ■

Example 3.4.1. We illustrate an application of Theorem 3.4.1(a) by constructing an $OA(27, 4, 3, 3)$. Take $m = g = 3$ in this theorem, and note that the elements of $GF(3)$ are $\rho_0 = 0$, $\rho_1 = 1$, $\rho_2 = 2$. Hence, as in the proof of Theorem 3.4.1(a), starting from

$$C = \begin{bmatrix} 1 & 1 & 1 & 0 \\ 0 & 1 & 2 & 0 \\ 0 & 1 & 1 & 1 \end{bmatrix}$$

we get an $OA(27, 4, 3, 3)$ as shown below:

$$\begin{bmatrix} 012 & 012 & 012 & 012 & 012 & 012 & 012 & 012 & 012 \\ 012 & 120 & 201 & 120 & 201 & 012 & 201 & 012 & 120 \\ 012 & 201 & 120 & 120 & 012 & 201 & 201 & 120 & 012 \\ 000 & 000 & 000 & 111 & 111 & 111 & 222 & 222 & 222 \end{bmatrix}'. \quad ■$$

Example 3.4.2. In Theorem 3.4.3, take $m = 3$, $r = 2$. Then, as in the proof of this theorem, with $T_1 = \{(1, 0)', (1, 1)', (1, 2)'\}$, $T_2 = \{(0, 1)'\}$ and

$$C = \begin{bmatrix} 1 & 1 & 1 & 0 \\ 0 & 1 & 2 & 1 \end{bmatrix},$$

an $OA(9, 4, 3, 2)$, isomorphic to the one shown in (2.6.2), can be obtained. As noted in Remark 3.4.2, such an orthogonal array could as well be constructed starting from a pair of MOLS of order 3:

$$\begin{bmatrix} 0 & 1 & 2 \\ 1 & 2 & 0 \\ 2 & 0 & 1 \end{bmatrix}, \quad \begin{bmatrix} 0 & 1 & 2 \\ 2 & 0 & 1 \\ 1 & 2 & 0 \end{bmatrix}. \qquad \blacksquare$$

Example 3.4.3. This example illustrates two more applications of Lemma 3.4.1 that are not covered by any of the theorems of this section.

(a) In Lemma 3.4.1 take $r = 4$, $n = 10$, $m = 3$, and

$$C = \begin{bmatrix} 0 & 0 & 0 & 0 & 1 & 1 & 1 & 1 & 1 & 1 \\ 0 & 0 & 1 & 1 & 0 & 0 & 1 & 1 & 2 & 2 \\ 1 & 0 & 1 & 2 & 1 & 2 & 1 & 2 & 1 & 2 \\ 0 & 1 & 1 & 2 & 1 & 2 & 2 & 1 & 2 & 1 \end{bmatrix}.$$

Then every 4×3 submatrix of C has rank 3 and an $OA(81, 10, 3, 3)$ can be constructed via Lemma 3.4.1. This construction is due to Bose and Bush (1952).

(b) In Lemma 3.4.1 take $r = 6$, $n = 8$, $m = 2$, and noting that the elements of $GF(2)$ are $\rho_0 = 0$, $\rho_1 = 1$, choose C as

$$\begin{bmatrix} 1 & 0 & 0 & 0 & 0 & 0 & 1 & 1 \\ 0 & 1 & 0 & 0 & 0 & 0 & 1 & 1 \\ 0 & 0 & 1 & 0 & 0 & 0 & 1 & 0 \\ 0 & 0 & 0 & 1 & 0 & 0 & 1 & 0 \\ 0 & 0 & 0 & 0 & 1 & 0 & 0 & 1 \\ 0 & 0 & 0 & 0 & 0 & 1 & 0 & 1 \end{bmatrix}.$$

Then every 6×4 submatrix of C has rank 4, and Lemma 3.4.1 yields an $OA(64, 8, 2, 4)$. ■

For some more related results, with a bearing on the construction of two-symbol orthogonal arrays of strength four or more, the reader may see Draper and Mitchell (1968) and the references therein.

3.5 METHOD OF DIFFERENCES

The method of differences is a powerful technique for the construction of orthogonal arrays. Bose and Bush (1952) initiated this method of construction. We begin with the definition of a *difference matrix*.

Definition 3.5.1. *Let $\lambda(\geq 1)$ and $m, n(\geq 2)$ be positive integers and $\mathcal{G}$ be a finite, additive Abelian group consisting of m elements. Then a difference matrix $D(\lambda m, n; m)$ is a $\lambda m \times n$ matrix with elements from $\mathcal{G}$ such that among the differences of the corresponding elements of every two distinct columns, each element of $\mathcal{G}$ appears λ times.*

For example,

$$D(6, 6; 3) = \begin{bmatrix} 0 & 1 & 1 & 0 & 2 & 2 \\ 2 & 2 & 1 & 0 & 0 & 1 \\ 2 & 1 & 2 & 0 & 1 & 0 \\ 0 & 0 & 0 & 0 & 0 & 0 \\ 1 & 2 & 0 & 0 & 1 & 2 \\ 1 & 0 & 2 & 0 & 2 & 1 \end{bmatrix} \tag{3.5.1}$$

is a difference matrix with elements from $GF(3)$.

It was shown by Jungnickel (1979) that in $D(\lambda m, n; m)$

$$n \leq \lambda m. \tag{3.5.2}$$

If the equality in (3.5.2) holds for a difference matrix, then it can be seen that its transpose is also a difference matrix. Such a matrix is called a generalized Hadamard matrix. Also note that the existence of a Hadamard matrix H_N of order $N(\geq 2)$ is equivalent to that of a difference matrix $D(N, N; 2)$ with elements from the additive group of integers reduced modulo 2. Replacing each -1 in H_N by 0, one gets $D(N, N; 2)$. Conversely, from $D(N, N; 2)$, one gets H_N by replacing each zero by -1. For a review on difference matrices, a reference may be made to de Launey (1986). Several useful difference matrices have been displayed, among others, by Masuyama (1957), Seiden (1954), and Wang and Wu (1991). See also the appendix in this connection.

We now have the following result:

Theorem 3.5.1. *(Bose and Bush, 1952) For positive integers $\lambda(\geq 1)$ and $m, n(\geq 2)$, the existence of a difference matrix $D(\lambda m, n; m)$, with elements from a finite, additive Abelian group $\mathcal{G}$ of order m, implies the existence of an orthogonal array $OA(\lambda m^2, n + 1, m, 2)$.*

Proof. Define the $\lambda m \times 1$ vector

$$\mathbf{e} = (e_0, \ldots, e_0, e_1, \ldots, e_1, \ldots, e_{m-1} \ldots, e_{m-1})',$$

where $e_0, e_1, \ldots, e_{m-1}$ are the elements of $\mathcal{G}$, each e_j being repeated λ times in $\mathbf{e}$. Also, denoting the difference matrix $D(\lambda m, n; m)$ by Δ, for $0 \leq j \leq$

$m - 1$, let Δ_j be a $\lambda m \times n$ matrix obtained by adding e_j to each element of Δ. Then it is easy to see that

$$\begin{bmatrix} \Delta_0' & \Delta_1' & \cdots & \Delta_{m-1}' \\ \mathbf{e}' & \mathbf{e}' & \cdots & \mathbf{e}' \end{bmatrix}'$$

is an $OA(\lambda m^2, n+1, m, 2)$. ■

Example 3.5.1. Starting from $D(6, 6; 3)$ shown in (3.5.1) and proceeding as in the proof of the above theorem, an $OA(18, 7, 3, 2)$, as exhibited below, can be constructed:

$$OA(18, 7, 3, 2) = \begin{bmatrix} 022 & 011 & 100 & 122 & 211 & 200 \\ 121 & 020 & 202 & 101 & 010 & 212 \\ 112 & 002 & 220 & 110 & 001 & 221 \\ 000 & 000 & 111 & 111 & 222 & 222 \\ 201 & 012 & 012 & 120 & 120 & 201 \\ 210 & 021 & 021 & 102 & 102 & 210 \\ 001 & 122 & 001 & 122 & 001 & 122 \end{bmatrix}'. \qquad ■$$

Remark 3.5.1. If m is a prime or prime power then $D(m, m; m)$ necessarily exists and is given by the multiplication table of $GF(m)$ so that by Theorem 3.5.1 one can construct an $OA(m^2, m+1, m, 2)$. The construction of such an orthogonal array, via mutually orthogonal Latin squares, was discussed earlier in Remark 3.4.2. ■

In view of Theorem 3.5.1, difference matrices play an important role in the construction of orthogonal arrays. The next two lemmas summarize some useful results on the existence and construction of difference matrices.

Lemma 3.5.1. *(Bose and Bush, 1952) If $\lambda(\geq 2)$ and $m(\geq 2)$ are powers of the same prime, then a difference matrix $D(\lambda m, \lambda m; m)$ can be constructed.*

Proof. Let $\lambda = p^u$ and $m = p^w$, where $p(\geq 2)$ is a prime and u, w are positive integers. Then $\lambda m = p^{u+w}$, and let the elements of $GF(\lambda m)$ be represented by polynomials (in y) of degree at most $u + w - 1$. Consider an additive Abelian group $\mathcal{G}$ of order m consisting of those $m(= p^w)$ elements of $GF(\lambda m)$ for which the coefficients of y^w and higher powers of y vanish. Now set up a correspondence, mapping ρ of $GF(\lambda m)$ to

$$e = a_{w-1}y^{w-1} + \cdots + a_1 y + a_0$$

of $\mathcal{G}$, the coefficients of y^{w-1} and lower powers of y in e being the same as the coefficients of the corresponding powers of y in the representation of ρ. Next, write down the multiplication table of $GF(\lambda m)$, and replace each element there by the corresponding element of $\mathcal{G}$. The resulting matrix is easily seen to be a difference matrix $D(\lambda m, \lambda m; m)$. ■

Example 3.5.2. This illustrates an application of Lemma 3.5.1 with $\lambda = 2$, $m = 4$. Then $p = 2$, $u = 1$, $w = 2$, $\lambda m = 8$, and the elements of $GF(8)$ can be expressed as polynomials in the primitive element y of $GF(8)$ as follows:

$$\rho_0 = 0,\ \rho_1 = 1,\ \rho_2 = y,\ \rho_3 = y^2,\ \rho_4 = y^3 = y + 1,$$
$$\rho_5 = y^4 = y^2 + y,\ \rho_6 = y^5 = y^2 + y + 1,\ \rho_7 = y^6 = y^2 + 1.$$

The elements of $\mathcal{G}$ are ρ_0, ρ_1, ρ_2, ρ_4, and the mapping of the elements of $GF(8)$ on the elements of $\mathcal{G}$ is given by

$$\rho_0, \rho_3 \to \rho_0, \quad \rho_1, \rho_7 \to \rho_1, \quad \rho_2, \rho_5 \to \rho_2, \quad \rho_4, \rho_6 \to \rho_4.$$

The multiplication table of $GF(8)$ is as shown below:

	ρ_0	ρ_1	ρ_2	ρ_3	ρ_4	ρ_5	ρ_6	ρ_7
ρ_0	ρ_0	ρ_0	ρ_0	ρ_0	ρ_0	ρ_0	ρ_0	ρ_0
ρ_1	ρ_0	ρ_1	ρ_2	ρ_3	ρ_4	ρ_5	ρ_6	ρ_7
ρ_2	ρ_0	ρ_2	ρ_3	ρ_4	ρ_5	ρ_6	ρ_7	ρ_1
ρ_3	ρ_0	ρ_3	ρ_4	ρ_5	ρ_6	ρ_7	ρ_1	ρ_2
ρ_4	ρ_0	ρ_4	ρ_5	ρ_6	ρ_7	ρ_1	ρ_2	ρ_3
ρ_5	ρ_0	ρ_5	ρ_6	ρ_7	ρ_1	ρ_2	ρ_3	ρ_4
ρ_6	ρ_0	ρ_6	ρ_7	ρ_1	ρ_2	ρ_3	ρ_4	ρ_5
ρ_7	ρ_0	ρ_7	ρ_1	ρ_2	ρ_3	ρ_4	ρ_5	ρ_6

By mapping the elements on $\mathcal{G}$, we get a difference matrix $D(8, 8; 4)$, displayed below:

$$D(8,8;4) = \begin{bmatrix} \rho_0 & \rho_0 & \rho_0 & \rho_0 & \rho_0 & \rho_0 & \rho_0 & \rho_0 \\ \rho_0 & \rho_1 & \rho_2 & \rho_0 & \rho_4 & \rho_2 & \rho_4 & \rho_1 \\ \rho_0 & \rho_2 & \rho_0 & \rho_4 & \rho_2 & \rho_4 & \rho_1 & \rho_1 \\ \rho_0 & \rho_0 & \rho_4 & \rho_2 & \rho_4 & \rho_1 & \rho_1 & \rho_2 \\ \rho_0 & \rho_4 & \rho_2 & \rho_4 & \rho_1 & \rho_1 & \rho_2 & \rho_0 \\ \rho_0 & \rho_2 & \rho_4 & \rho_1 & \rho_1 & \rho_2 & \rho_0 & \rho_4 \\ \rho_0 & \rho_4 & \rho_1 & \rho_1 & \rho_2 & \rho_0 & \rho_4 & \rho_2 \\ \rho_0 & \rho_1 & \rho_1 & \rho_2 & \rho_0 & \rho_4 & \rho_2 & \rho_4 \end{bmatrix}.$$

Now, by Theorem 3.5.1, an $OA(32, 9, 4, 2)$ can be obtained. ■

Lemma 3.5.2. *(Street, 1979; Jungnickel, 1979) If $m(\geq 2)$ is a prime or prime power, then a difference matrix $D(2m, 2m; m)$ can be constructed.*

Proof.

(a) If m is even then $m = 2^r$ for some positive integer r and the result follows from Lemma 3.5.1.

(b) Now suppose that m is an odd prime or prime power. We sketch a proof following Dawson (1985). For $i, j = 1, 2$, let α_{ij}, β_{ij}, γ_{ij} be elements of $GF(m)$ such that

$$\frac{\alpha_{11} - \alpha_{21}}{\beta_{11}\beta_{21}} = \frac{\alpha_{12} - \alpha_{22}}{\beta_{12}\beta_{22}}, \tag{3.5.3}$$

$$\frac{\gamma_{11} - \gamma_{12}}{\beta_{11}\beta_{12}} = \frac{\gamma_{21} - \gamma_{22}}{\beta_{21}\beta_{22}}, \tag{3.5.4}$$

$$4\left(\frac{\alpha_{11} - \alpha_{21}}{\beta_{11}\beta_{21}}\right)\left(\frac{\gamma_{11} - \gamma_{12}}{\beta_{11}\beta_{22}}\right) = \frac{\beta_{11}\beta_{22} - \beta_{12}\beta_{21}}{\beta_{11}\beta_{12}\beta_{21}\beta_{22}}, \tag{3.5.5}$$

and

$$\beta_{11}\beta_{12}\beta_{21}\beta_{22} \quad \text{is a nonquadratic residue of } GF(m), \tag{3.5.6}$$

the numerators and denominators in (3.5.3)–(3.5.5) being nonzero. Note that for odd m, it is always possible to find elements of $GF(m)$ satisfying (3.5.3)–(3.5.6): First choose the β_{ij}'s satisfying (3.5.6), then choose $\alpha_{11}, \alpha_{21}, \gamma_{11}$ and γ_{12} satisfying (3.5.5) and finally choose α_{12}, α_{22}, γ_{21} and γ_{22} satisfying (3.5.3) and (3.5.4). Denoting the elements of $GF(m)$ by $\rho_0, \rho_1, \ldots, \rho_{m-1}$, for $i, j = 1, 2$, now let D_{ij} be an $m \times m$ matrix with (u, w)th element $\alpha_{ij}\rho_w{}^2 + \beta_{ij}\rho_u\rho_w + \gamma_{ij}\rho_u{}^2$, $0 \leq u, w \leq m - 1$. Then, as detailed in Dawson (1985),

$$\begin{bmatrix} D_{11} & D_{12} \\ D_{21} & D_{22} \end{bmatrix}'$$

represents a difference matrix $D(2m, 2m; m)$. ∎

Example 3.5.3. In the setup of Lemma 3.5.2, let $m = 3$. The elements of $GF(3)$ are $\rho_0 = 0$, $\rho_1 = 1$, $\rho_2 = 2$, and it can be seen that if

$$\alpha_{11} = 2, \alpha_{12} = 1, \alpha_{21} = \alpha_{22} = 0,$$

$$\gamma_{11} = 1, \gamma_{12} = 0, \gamma_{21} = 2, \gamma_{22} = 0,$$

$$\beta_{11} = 2, \beta_{12} = \beta_{21} = \beta_{22} = 1,$$

then (3.5.3)–(3.5.6) are satisfied. Hence, as in part (b) of the proof of Lemma 3.5.2, one can construct $D(6, 6; 3)$. This will be as shown in (3.5.1). ■

The following corollaries are evident from Theorem 3.5.1 and Lemmas 3.5.1 and 3.5.2:

Corollary 3.5.1. *Let $\lambda(\geq 2)$ and $m(\geq 2)$ be powers of the same prime. Then an orthogonal array $OA(\lambda m^2, \lambda m + 1, m, 2)$ can be constructed.* ■

Corollary 3.5.2. *If $m(\geq 2)$ is a prime or prime power, then an orthogonal array $OA(2m^2, 2m + 1, m, 2)$ can be constructed.* ■

Remark 3.5.2. As will be seen in Chapter 5, in an $OA(2m^2, n, m, 2)$, if $m \geq 3$, then n cannot exceed $2m + 1$. The constructions in Examples 3.5.1 and 3.5.2 lead to the attainment of this upper bound on n. More generally, for $m \geq 3$, this upper bound is attained by orthogonal arrays constructed via Corollary 3.5.2. ■

Remark 3.5.3. Dawson (1985) also proved the existence of a difference matrix $D(4m, 4m; m)$ and hence of an orthogonal array $OA(4m^2, 4m+1, m, 2)$ when m is a prime or a prime power. For $m = 2^r$, the existence of such a difference matrix is evident from Lemma 3.5.1. For odd m, the proof of the result is much more involved and we refer to Dawson (1985) for the details. ■

Remark 3.5.4. Addelman and Kempthorne (1961b) described an ingenious procedure for constructing a series of arrays $OA(2m^r, 2(m^r - 1)/(m - 1) - 1, m, 2)$, where $m(\geq 2)$ is a prime or prime power and $r(\geq 2)$ is an integer. In the special case $r = 2$, their approach yields an $OA(2m^2, 2m + 1, m, 2)$, which has the same parameters as (but is not necessarily isomorphic to) an orthogonal array arising from Corollary 3.5.2 by the method of differences. The interested reader is referred to Addelman and Kempthorne (1961b) for more details on their method of construction. ■

Remark 3.5.5. Hedayat et al. (1996) generalized the notion of difference matrices to define difference schemes, investigated the problems of existence and construction of such schemes, and explored their use in the construction of orthogonal arrays of strength greater than two. We refer to the original paper for the details. ■

3.6 SOME FURTHER RESULTS

Mukhopadhyay (1981) obtained several series of orthogonal arrays of strength two and three. These results are summarized below.

Theorem 3.6.1. *Suppose that m and $m-1$ are both primes or prime powers, and that $n \geq 2$ and $u \geq 1$ are integers. Then there exist orthogonal arrays*

(a) $OA((m-1)^u m^{n+u}, (m^{n-1}+m^{n-2}+\cdots+m+1)(m-1)^{2u}+(m-1)^{2(u-1)}+\cdots+1, m, 2)$;

(b) $OA(2(m-1)^u m^{n+u}, \{2(m^{n-1}+m^{n-2}+\cdots+m)+1\}(m-1)^{2u}+(m-1)^{2(u-1)}+\cdots+1, m, 2)$. ■

Theorem 3.6.2. *If λ and m are powers of the same prime, then there exists an orthogonal array $OA(\lambda^{g-1}m^g, \lambda m, m, g)$, $g \geq 3$.* ■

We refer to Mukhopadhyay (1981) for constructive proofs of the above theorems. Further results in continuation of the above were reported by Goswami and Pal (1992).

Given orthogonal arrays $OA(N_i, n_i, m_i, g)$, $1 \leq i \leq r$, one can get another orthogonal array, by taking "product" of these arrays. The following result was proved by Bush (1952) in this context.

Theorem 3.6.3. *The existence of arrays $OA(N_i, n_i, m_i, g)$, $1 \leq i \leq r$, implies the existence of an orthogonal array $OA(N, n, m, g)$, where $N = \prod N_i$, $m = \prod m_i$ and $n = min(n_1, \ldots, n_r)$.*

Proof. Let $A = ((a_{ij}))$ denote the transpose of the array $OA(N_1, n_1, m_1, g)$ and $B = ((b_{ij}))$, the transpose of the array $OA(N_2, n_2, m_2, g)$. Let $n = min(n_1, n_2)$. Form the $n \times N_1N_2$ array

$$\begin{array}{ccccccc}
(a_{11}b_{11}) & \cdots & (a_{11}b_{1N_2}) & \cdots & (a_{1N_1}b_{11}) & \cdots & (a_{1N_1}b_{1N_2}), \\
(a_{21}b_{21}) & \cdots & (a_{21}b_{2N_2}) & \cdots & (a_{2N_1}b_{21}) & \cdots & (a_{2N_1}b_{2N_2}), \\
\vdots & & & & & & \\
(a_{n1}b_{n1}) & \cdots & (a_{n1}b_{nN_2}) & \cdots & (a_{nN_1}b_{n1}) & \cdots & (a_{nN_1}b_{nN_2}),
\end{array}$$

whose elements are ordered pairs. The transpose of the above array can be verified to be an orthogonal array $OA(N_1N_2, n, m_1m_2, g)$. By repeated use of this fact, we get the required result. ■

Several results on *embedding* of orthogonal arrays are available in the literature. By embedding we mean any procedure of adding more columns to an orthogonal array without increasing the number of rows or reducing the strength. The interested reader is referred to Seiden (1954), Seiden and Zemach (1966) and Shrikhande and Bhagwandas (1969, 1970) for details in this regard.

EXERCISES

3.1. In (3.3.1), let A be an $OA(N, n, 2, 2s - 1)$ of odd strength and $\bar{A}$ be obtained by interchanging the symbols in A. Then is it true that A_1 will necessarily be an $OA(2N, n + 1, 2, 2s)$?

3.2. Consider the following conjecture which reduces to Theorem 3.3.1 for $m = 2$ and, if true, gives a generalization of this theorem:

> For integers $m(\geq 2)$ and $s(\geq 1)$, the existence of an $OA(N, n, m, 2s)$ is equivalent to that of an $OA(mN, n + 1, m, 2s + 1)$.

Show that the conjecture is not true for general m. [*Hint*: See Remarks 3.4.1 and 3.4.2.]

3.3. Let $m(\geq 4)$ be an even prime power. Then verify that every 3×3 submatrix of the matrix C considered in the proof of Theorem 3.4.1(b) has full column rank. Also check that this is not the case if m happens to be an odd prime or prime power.

3.4. Verify that every 6×4 submatrix of the matrix C considered in Example 3.4.3(b) has full column rank.

3.5. Show that in contrast with Remark 3.5.1, the addition table of $GF(m)$, where m is a prime or a prime power, does not represent a difference matrix $D(m, m; m)$.

3.6. Supply all details underlying the construction in Lemma 3.5.1.

3.7. Use Lemma 3.5.2 to construct a difference matrix $D(10, 10; 5)$.

CHAPTER 4

Asymmetric Orthogonal Arrays

4.1 INTRODUCTION

In Chapter 3 methods of construction of *symmetric* orthogonal arrays have been described. In this chapter we discuss several procedures for constructing asymmetric orthogonal arrays. Methods of construction of asymmetric arrays of strength *two* are reviewed in Sections 4.2 through 4.6. Section 4.2 deals with the methods of collapsing and replacement. The methods of construction based on Hadamard matrices are treated in Section 4.3. Difference matrices and Kronecker sum of matrices are useful in generating a large number of asymmetric arrays of strength two. These methods are discussed in Section 4.4. In Section 4.5 we indicate the role of resolvable orthogonal arrays in the construction of some families of asymmetric orthogonal arrays of strength two. A method of grouping is briefly reviewed in Section 4.6. In Section 4.7 orthogonal arrays of higher strength are considered, and the emphasis is on arrays of strength three.

4.2 COLLAPSING AND REPLACEMENT PROCEDURES

The methods of collapsing and replacement are due to Addelman (1962a). Suppose that an orthogonal array of strength two has a column involving m symbols, and let $m'(\geq 2)$ be a positive integer such that $m' \mid m$, namely, $m = 0 \pmod{m'}$. Then the m-symbol column can be "collapsed" into an m'-symbol column by first grouping the m symbols into m' sets of m/m' symbols each and then replacing the symbols belonging to the same set by a common symbol. It is easy to see that the resulting array is still an orthogonal array of strength two. We illustrate the method below.

Example 4.2.1. Consider the array $OA(16, 5, 4, 2)$, shown below, which can be constructed as in Remark 3.4.2:

$$OA(16,5,4,2) = \begin{bmatrix} 0000 & 1111 & 2222 & 3333 \\ 0123 & 0123 & 0123 & 0123 \\ 0123 & 1032 & 2301 & 3210 \\ 0123 & 2301 & 3210 & 1032 \\ 0123 & 3210 & 1032 & 2301 \end{bmatrix}'.$$

Since $m' = 2$ divides $m = 4$, we group the four symbols of the first column, say, into two sets, say, $\{0, 1\}$ and $\{2, 3\}$. Replacing both the symbols in the first set by 0 and those in the second set by 1, we get an asymmetric orthogonal array $OA(16, 5, 2 \times 4^4, 2)$ (in the notation of Chapter 2), which is displayed below:

$$OA(16,5,2\times 4^4,2) = \begin{bmatrix} 0000 & 0000 & 1111 & 1111 \\ 0123 & 0123 & 0123 & 0123 \\ 0123 & 1032 & 2301 & 3210 \\ 0123 & 2301 & 3210 & 1032 \\ 0123 & 3210 & 1032 & 2301 \end{bmatrix}'.$$ ■

Orthogonal arrays of strength two obtained through the procedure of collapsing can be found in the catalog of Addelman and Kempthorne (1961a) and in Dey (1985).

Next, we take up the procedure of "replacement." Suppose that an orthogonal array A of strength two has a column involving m symbols, and let there exist another orthogonal array B, also of strength two in $N = m$ rows. Then in A the symbols of the m-symbol column can be replaced by the rows of B, using one–one correspondence, without disturbing orthogonality. From Remarks 2.6.3 and 2.6.4 it is easily seen that if A and B are both tight, then this method of replacement yields a tight orthogonal array of strength two. The method of replacement reduces to that of collapsing in the special case where B consists of a single column and is therefore more powerful than the latter in the sense of yielding orthogonal arrays with more columns.

Example 4.2.2. Let A represent the $OA(16, 5, 4, 2)$ shown in Example 4.2.1. Replacing the symbols 0, 1, 2, 3 in the first column of A, respectively, by the rows 000, 011, 101, 110 of an $OA(4, 3, 2, 2)$, obtainable via Theorem 3.2.2, we get a tight asymmetric $OA(16, 7, 2^3 \times 4^4, 2)$. This is better than the $OA(16, 5, 2 \times 4^4, 2)$, obtained in Example 4.2.1 in the sense of having two additional two-symbol columns. ■

Using the replacement procedure, an orthogonal array of strength two with one or more columns having a large number of symbols can be converted into an orthogonal array of strength two with columns having relatively smaller number of symbols. Many such orthogonal arrays are displayed in the catalog of Addelman and Kempthorne (1961a).

4.3 USE OF HADAMARD MATRICES

Several asymmetric orthogonal arrays of strength two can be constructed by an intelligent use of the properties of Hadamard matrices, as demonstrated by Dey and Ramakrishna (1977), Chacko, Dey, and Ramakrishna (1979), Chacko and Dey (1981), Agrawal and Dey (1982), Cheng (1980b, 1989), and Wang (1990). We describe these construction methods in this section. The results due to Cheng (1989) are presented first since they generalize and unify the earlier findings reported by Dey and others.

The following preliminaries will be helpful: A positive integer N will be called a *Hadamard number* if a Hadamard matrix, H_N, of order N exists; throughout, unless otherwise stated, the trivial Hadamard number $N = 1$ will be left out of consideration. As indicated in the last chapter, without loss of generality, the initial column of H_N will be supposed to be $\mathbf{1}_N$, the $N \times 1$ vector of all ones. A set of three distinct columns of $H_N (N \geq 4)$ is said to have the *Hadamard property* if the Hadamard product of any two columns in the set equals the third (recall that the Hadamard product of two vectors $\mathbf{a} = (a_1, \ldots, a_v)'$ and $\mathbf{b} = (b_1, \ldots, b_v)'$ is defined as $\mathbf{a} * \mathbf{b} = (a_1 b_1, \ldots, a_v b_v)'$). Evidently the $N \times 3$ submatrix of H_N given by three such columns has rows $(1, 1, 1)$, $(1, -1, -1)$, $(-1, 1, -1)$ and $(-1, -1, 1)$ each with frequency $N/4$. We now have the following result, stated by Wang (1990):

Lemma 4.3.1. *Let H_N be a Hadamard matrix of order $N(\geq 8)$, and suppose that there is a set of three columns of H_N having the Hadamard property. Denote these columns by $\boldsymbol{a}_1$, $\boldsymbol{a}_2$, $\boldsymbol{a}_3$. Then in any $N \times 4$ submatrix of H_N made up of the columns $\boldsymbol{a}_i$, $i = 1, 2, 3$, and $\mathbf{c}$, where $\mathbf{c}$ is any column of H_N other than $\mathbf{1}_N$ and the three with the Hadamard property, each of the 8 vectors $(1, 1, 1, \pm 1)$, $(1, -1, -1, \pm 1)$, $(-1, 1, -1, \pm 1)$, $(-1, -1, 1, \pm 1)$ occurs equally often as a row.*

Proof. Let the frequencies of the vectors $(1, 1, 1, 1)$, $(1, -1, -1, 1)$, $(-1, 1, -1, 1)$, and $(-1, -1, 1, 1)$, as rows of the $N \times 4$ submatrix under consideration, be x_1, x_2, x_3, x_4, respectively. As each of $(1, 1, 1)$, $(1, -1, -1)$, $(-1, 1, -1)$, and $(-1, -1, 1)$ appears $N/4$ times as a row of the $N \times 3$ submatrix given by $\boldsymbol{a}_1$, $\boldsymbol{a}_2$, $\boldsymbol{a}_3$, it is enough to show that $x_1 = x_2 = x_3 = x_4 = N/8$.

Since $\boldsymbol{c}$ is orthogonal to each of $\boldsymbol{a}_1$, $\boldsymbol{a}_2$, and $\boldsymbol{a}_3$, it follows from the orthogonality of $\boldsymbol{c}$ and $\boldsymbol{a}_1$ that

$$\left\{x_1 - \left(\frac{N}{4} - x_1\right)\right\} + \left\{x_2 - \left(\frac{N}{4} - x_2\right)\right\} - \left\{x_3 - \left(\frac{N}{4} - x_3\right)\right\} - \left\{x_4 - \left(\frac{N}{4} - x_4\right)\right\} = 0,$$

namely,

$$x_1 + x_2 - x_3 - x_4 = 0. \tag{4.3.1}$$

Similarly, invoking the orthgonality of $\boldsymbol{c}$ and each of $\boldsymbol{a}_2$ and $\boldsymbol{a}_3$, we have

$$x_1 - x_2 + x_3 - x_4 = 0, \tag{4.3.2}$$

$$x_1 - x_2 - x_3 + x_4 = 0. \tag{4.3.3}$$

Also, since $\boldsymbol{c}$ is different from $\mathbf{1}_N$,

$$x_1 + x_2 + x_3 + x_4 = \frac{N}{2}. \tag{4.3.4}$$

The unique solution of the equations (4.3.1)–(4.3.4) is $x_1 = x_2 = x_3 = x_4 = N/8$. ■

Suppose that there is a Hadamard matrix H_N containing a set of three columns with the Hadamard property. If one replaces the four rows $(1, 1, 1)$, $(1, -1, -1)$, $(-1, 1, -1)$, $(-1, -1, 1)$ under the columns with the Hadamard property by the symbols 0, 1, 2, 3, respectively, and also deletes the initial column $\mathbf{1}_N$ of H_N, then the resulting $N \times (N - 3)$ matrix, by virtue of Lemma 4.3.1, is easily seen to represent a tight array $OA(N, N - 3, 4 \times 2^{N-4}, 2)$. More generally, if H_N includes $t (\leq (N - 1)/3)$ disjoint sets of columns such that each set has the Hadamard property, then the same technique of deleting $\mathbf{1}_N$ and replacing each set of three columns by a single four-symbol column yields a tight orthogonal array $OA(N, N - 2t - 1, 4^t \times 2^{N-3t-1}, 2)$.

We are now in a position to present one of Cheng's (1989) results which was also proved by Wang (1990). We follow Wang's proof here.

Theorem 4.3.1. *If N, T are Hadamard numbers satisfying $N \geq 4$, $T \leq N$, then there exists an orthogonal array $OA(NT, NT - 2T + 1, 4^{T-1} \times 2^{NT-3T+2}, 2)$, which is tight.*

Proof. Write

$$H_N = [\mathbf{1}_N \vdots \mathbf{b}_1 \vdots \cdots \vdots \mathbf{b}_{N-1}],\ H_T = [\mathbf{1}_T \vdots \mathbf{a}_1 \vdots \cdots \vdots \mathbf{a}_{T-1}], \tag{4.3.5}$$

where $\mathbf{b}_1, \ldots, \mathbf{b}_{N-1}$ are the columns of H_N other than the initial column and similarly $\mathbf{a}_1, \ldots, \mathbf{a}_{T-1}$ are defined. Considering $H_{NT} = H_T \otimes H_N$, it is not hard to see that H_{NT} contains $T-1$ disjoint sets of columns, given by $\{\mathbf{1}_T \otimes \mathbf{b}_i, \mathbf{a}_i \otimes \mathbf{1}_N, \mathbf{a}_i \otimes \mathbf{b}_i\}$, $1 \le i \le T-1$, each set having the Hadamard property. Hence, as indicated above, the required orthogonal array can be constructed.

■

Remark 4.3.1. Taking $T = 2, 4$ in Theorem 4.3.1, we get the arrays $OA(2N, 2N-3, 4 \times 2^{2N-4}, 2)$ and $OA(4N, 4N-7, 4^3 \times 2^{4N-10}, 2)$, respectively. These arrays were constructed earlier by Dey and Ramakrishna (1977) and Chacko, Dey, and Ramakrishna (1979), respectively. For larger values of T, other arrays are obtained. For example, for $T = 8$, $N = 12$, one obtains the tight array $OA(96, 81, 4^7 \times 2^{74}, 2)$. ■

Example 4.3.1. Let $N = 4$, $T = 2$,

$$H_4 = H_2 \otimes H_2 = \begin{bmatrix} 1 & 1 & 1 & 1 \\ 1 & -1 & 1 & -1 \\ 1 & 1 & -1 & -1 \\ 1 & -1 & -1 & 1 \end{bmatrix},$$

$$H_2 = \begin{bmatrix} 1 & 1 \\ 1 & -1 \end{bmatrix}.$$

Then, following (4.3.5), $\mathbf{a}_1 = (1, -1)'$, $\mathbf{b}_1 = (1, -1, 1, -1)'$. Now, as in the proof of Theorem 4.3.1, if one forms $H_8 = H_2 \otimes H_4$, replaces the triplet of columns$\{\mathbf{1}_2 \otimes \mathbf{b}_1, \mathbf{a}_1 \otimes \mathbf{1}_4, \mathbf{a}_1 \otimes \mathbf{b}_1\}$ there by a single four-symbol column, and finally deletes the initial column of all ones, then one obtains the array

$$OA(8, 5, 4 \times 2^4, 2) = \begin{bmatrix} 0 & 1 & 1 & 1 & 1 \\ 2 & 1 & -1 & 1 & -1 \\ 0 & -1 & -1 & -1 & -1 \\ 2 & -1 & 1 & -1 & 1 \\ 1 & 1 & 1 & -1 & -1 \\ 3 & 1 & -1 & -1 & 1 \\ 1 & -1 & -1 & 1 & 1 \\ 3 & -1 & 1 & 1 & -1 \end{bmatrix},$$

which is isomorphic to the one shown in (2.6.1). ■

Cheng (1989) also proved Theorem 4.3.2 below using the properties of Hadamard matrices. An alternative proof of this theorem will be indicated in the next section (see Remark 4.4.2).

Theorem 4.3.2. *If N and $T/2$ are Hadamard numbers, then there exists a tight orthogonal array $OA(NT, N(T-1)-2s+1, N \times 4^s \times 2^{N(T-1)-3s}, 2)$, where $s = min(N-1, T-1)$.* ■

Wang (1990) reported another proof of Theorem 4.3.2 for the case $N \geq T$. We refer to Wang's (1990) paper for this and other details.

Remark 4.3.2. In Theorem 4.3.2, let $T = 4$. Then, for each Hadamard number $N \geq 4$, we have the array $OA(4N, 3N-5, N \times 4^3 \times 2^{3N-9}, 2)$, which was constructed earlier by Agrawal and Dey (1982). ■

Remark 4.3.3. The idea of replacing a set of three columns having the Hadamard property by a four-symbol column can be extended to the case of $2^\nu - 1$ columns ($\nu \geq 2$) with a similar property. For instance, with $\nu = 3$, if we can find seven distinct columns $\mathbf{a}_1, \mathbf{a}_2, \ldots, \mathbf{a}_7$ of a Hadamard matrix H_N such that $\mathbf{a}_1 * \mathbf{a}_2 = \mathbf{a}_3$, $\mathbf{a}_1 * \mathbf{a}_4 = \mathbf{a}_5$, $\mathbf{a}_2 * \mathbf{a}_4 = \mathbf{a}_6$ and $\mathbf{a}_3 * \mathbf{a}_4 = \mathbf{a}_7$, then the submatrix given by these seven columns will have eight distinct rows, each repeated $N/8$ times. As before, these eight rows can be replaced by eight distinct symbols; namely, the columns $\mathbf{a}_1, \mathbf{a}_2, \ldots, \mathbf{a}_7$ can be replaced by a single eight-symbol column retaining orthogonality. ■

Generalizing Theorem 4.3.1, Dey (1993) obtained the following result:

Theorem 4.3.3. *Let $N(\geq 4)$ and T be Hadamard numbers. Then the existence of a symmetric orthogonal array $OA(T, k, u, 2)$ such that $k \mid (T-1)$ and $k \leq N-1$ implies the existence of an orthogonal array $OA(NT, NT - 2T + 1, (2u)^k \times 2^{NT-2T-k+1}, 2)$.*

Proof. Let the array $OA(T, k, u, 2)$ exist, and let $k \mid (T-1)$. Without loss of generality, let the symbols of this array be $0, 1, \ldots, u-1$. Write H_N as

$$H_N = [\mathbf{1}_N \vdots \mathbf{b}_1 \vdots \cdots \vdots \mathbf{b}_k \vdots B],$$

where for $i = 1, 2, \ldots, k$, $\mathbf{b}_i$'s are any k columns of H_N (other than the initial column) and B is the $N \times (N-k-1)$ matrix of remaining columns. For $i = 1, 2, \ldots, k$, replace in the ith column of $OA(T, k, u, 2)$, the symbol j by $(2j+1)\mathbf{b}_i$, $j = 0, 1, \ldots, u-1$. Call the resultant matrix D_1. Each of the k columns of D_1 involves $2u$ distinct symbols, coded as $\pm 1, \pm 3, \cdots,$

$\pm(2u-1)$. Next write H_T as

$$H_T = [\mathbf{1}_T \vdots \mathbf{a}_1 \vdots \mathbf{a}_2 \vdots \cdots \vdots \mathbf{a}_{mk}],$$

where $m = (T-1)/k$. For $i = 1, 2, \ldots, k$, define C_i to be the $N \times (k-1)$ matrix obtained by deleting $\mathbf{b}_i$ from $[\mathbf{b}_1 \vdots \mathbf{b}_2 \vdots \cdots \vdots \mathbf{b}_k]$, and let

$$D_2{}^* = [C_1{}^* \vdots C_2{}^* \vdots \cdots \vdots C_m{}^*],$$

where, for $i = 1, \ldots, m$,

$$C_i{}^* = [\mathbf{a}_{(i-1)k+1} \otimes C_1 \vdots \mathbf{a}_{(i-1)k+2} \otimes C_2 \vdots \cdots \vdots \mathbf{a}_{ik} \otimes C_k].$$

Finally, define $D_2 = [D_2{}^* \vdots H_T \otimes B]$. Then it can be verified that $D = [D_1 \vdots D_2]$ gives the required orthogonal array. ■

Remark 4.3.4. It is easily seen that Theorem 4.3.3 yields a tight orthogonal array, provided the array $OA(T, k, u, 2)$ is tight. Some special cases are of interest. Since T is a Hadamard number, there exists an array $OA(T, T-1, 2, 2)$ (*vide* Theorem 3.2.2) satisfying the conditions of Theorem 4.3.3. Hence for a Hadamard number $N \geq 4$, T, we have the array $OA(NT, NT-2T+1, 4^{T-1} \times 2^{NT-3T+2}, 2)$, obtained earlier via Theorem 4.3.1. Again, if $u = 2^v$ for some positive integer v, then u^2 is a Hadamard number and the array $OA(u^2, u+1, 2, 2)$ exists (see Remark 3.4.2). Hence, from Theorem 4.3.3, for every Hadamard number $N \geq u+2$, we have the array $OA(Nu^2, Nu^2-2u^2+1, (2u)^{u+1} \times 2^{Nu^2-2u^2-u}, 2)$, which is tight. ■

As another application of Hadamard matrices, we have the following result due to Agrawal and Dey (1982):

Theorem 4.3.4. *If $N(\geq 4)$ is a Hadamard number, then the array $OA(2N, N+1, T \times 4 \times 2^{N-1}, 2)$ exists, where $T = N/2$.*

Proof. Let $H_N = [\mathbf{1}_N \vdots \mathbf{b}_1 \vdots B_2]$, where B_2 is $N \times (N-2)$. Without loss of generality (by a permutation of rows if necessary), let the first $N/2$ elements of $\mathbf{b}_1$ equal $+1$ and the remaining elements of $\mathbf{b}_1$ equal -1. Let $\mathbf{c}_1 = 3\mathbf{b}_1$, $\mathbf{a}_1 = (1, 2, \ldots, T, 1, 2, \ldots, T)'$ and $\mathbf{a}_2$ represent an $N \times 1$ vector with elements -1 in the first and last $N/4$ positions and $+1$ in the middle $N/2$

positions. Then it can be verified that

$$D = \begin{bmatrix} \mathbf{a}_1 & \mathbf{b}_1 & B_2 & \mathbf{a}_2 \\ \mathbf{a}_1 & \mathbf{c}_1 & -B_2 & \mathbf{a}_2 \end{bmatrix}$$

gives the required orthogonal array. ■

Example 4.3.2. Let $N = 12$ in Theorem 4.3.4. Then, starting from

$$H_{12} = \begin{bmatrix} 1 & 1 & 1 & 1 & 1 & 1 & 1 & 1 & 1 & 1 & 1 & 1 \\ 1 & 1 & -1 & 1 & 1 & 1 & -1 & -1 & -1 & 1 & -1 & -1 \\ 1 & 1 & -1 & -1 & 1 & -1 & 1 & 1 & 1 & -1 & -1 & -1 \\ 1 & 1 & -1 & -1 & -1 & 1 & -1 & -1 & 1 & -1 & 1 & 1 \\ 1 & 1 & 1 & -1 & -1 & -1 & 1 & -1 & -1 & 1 & -1 & 1 \\ 1 & 1 & 1 & 1 & -1 & -1 & -1 & 1 & -1 & -1 & 1 & -1 \\ 1 & -1 & 1 & -1 & 1 & 1 & 1 & -1 & -1 & -1 & 1 & -1 \\ 1 & -1 & -1 & 1 & -1 & 1 & 1 & 1 & -1 & -1 & -1 & 1 \\ 1 & -1 & 1 & -1 & -1 & 1 & -1 & 1 & 1 & 1 & -1 & -1 \\ 1 & -1 & -1 & 1 & -1 & -1 & 1 & -1 & 1 & 1 & 1 & -1 \\ 1 & -1 & -1 & -1 & 1 & -1 & -1 & 1 & -1 & 1 & 1 & 1 \\ 1 & -1 & 1 & 1 & 1 & -1 & -1 & -1 & 1 & -1 & -1 & 1 \end{bmatrix},$$

one gets the array $OA(24, 13, 6 \times 4 \times 2^{11}, 2)$, shown below. Here, for simplicity, in the second column of the array, the symbols are coded using the transformation $-3 \rightarrow 0$, $-1 \rightarrow 1$, $1 \rightarrow 2$, $3 \rightarrow 3$. Similarly, in the two-symbol columns, -1 is replaced by 0.

$$OA(24, 13, 6 \times 4 \times 2^{11}, 2) = \begin{bmatrix} 123456 & 123456 & 123456 & 123456 \\ 222222 & 111111 & 333333 & 000000 \\ 100011 & 101001 & 011100 & 010110 \\ 110001 & 010101 & 001110 & 101010 \\ 111000 & 100011 & 000111 & 011100 \\ 110100 & 111000 & 001011 & 000111 \\ 101010 & 110100 & 010101 & 001011 \\ 101001 & 011010 & 010110 & 100101 \\ 101100 & 001101 & 010011 & 110010 \\ 110010 & 001110 & 001101 & 110001 \\ 100101 & 100110 & 011010 & 011001 \\ 100110 & 010011 & 011001 & 101100 \\ 000111 & 111000 & 000111 & 111000 \end{bmatrix}'. \quad ■$$

4.4 USE OF DIFFERENCE MATRICES

The use of difference matrices for the construction of asymmetric orthogonal arrays of strength two was first made by Suen (1989a). Wang and Wu (1991) suggested a somewhat unified method of constructing asymmetric orthogonal arrays of strength two based on difference matrices and *Kronecker sum* of matrices. This method will be described in the present section. Difference matrices have already been introduced in Chapter 3 (see Definition 3.5.1). We now have the definition of Kronecker sum of two matrices.

Definition 4.4.1. *Let A and B be two matrices of orders $n \times r$ and $m \times s$, respectively, and let A and B have entries from a finite additive Abelian group $\mathcal{G}$ of order p. The Kronecker sum of A and B, denoted by $A \odot B$, is defined to be the partitioned matrix*

$$A \odot B = (B^{a_{ij}})_{1 \le i \le n, 1 \le j \le r},$$

where $A = (a_{ij})$ and $B^{a_{ij}}$ is an $m \times s$ matrix obtained by adding a_{ij} to each element of B.

Suppose that $A = [A_1 \vdots A_2 \vdots \cdots \vdots A_n]$ and $B = [B_1 \vdots B_2 \vdots \cdots \vdots B_n]$ are two partitioned matrices, such that for each i, both A_i and B_i have entries from a finite additive Abelian group $\mathcal{G}_i$ ($i = 1, 2, \ldots, n$). The generalized Kronecker sum of A and B, denoted by $A \oplus B$, is defined as

$$A \oplus B = [A_1 \odot B_1 \vdots A_2 \odot B_2 \vdots \cdots \vdots A_n \odot B_n].$$

Wang and Wu (1991) proved the following basic result.

Theorem 4.4.1. *Let A be an orthogonal array $OA(N, n_1 + n_2 + \cdots + n_u, {m_1}^{n_1} \times {m_2}^{n_2} \times \cdots \times {m_u}^{n_u}, 2)$, and suppose that there exist difference matrices, $D(M, k_i; m_i)$, $i = 1, 2, \ldots, u$, where N and M are both multiples of the m_i's. Partition A as $A = [C_1 \vdots C_2 \vdots \cdots \vdots C_u]$, where for $i = 1, 2, \ldots, u$, C_i is a symmetric orthogonal array $OA(N, n_i, m_i, 2)$, with symbols from the Abelian group over which $D(M, k_i; m_i)$ is defined. Then the generalized Kronecker sum*

$$[C_1 \odot D(M, k_1; m_1) \vdots C_2 \odot D(M, k_2; m_2) \vdots \cdots \vdots C_u \odot D(M, k_u; m_u)]$$

is an orthogonal array $OA(MN, \sum_{i=1}^{u} k_i n_i, {m_1}^{k_1 n_1} \times {m_2}^{k_2 n_2} \times \cdots \times {m_u}^{k_u n_u}, 2)$.

Proof. Let $E_i = C_i \odot D(M, k_i; m_i)$, and let us, for notational convenience, denote an $OA(N, n, m, 2)$ by $L_N(m^n)$. A result of Shrikhande (1964) states that if there exist an orthogonal array $L_{\mu m}(m^n)$ and a difference matrix $D(\lambda m, r; m)$, then

$$D = L_{\mu m}(m^n) \odot D(\lambda m, r; m)$$

is an orthogonal array $L_{\lambda\mu m^2}(m^{rn})$. By this result each E_i, $i = 1, 2, \ldots, u$ is an orthogonal array $L_{MN}(m_i^{k_i n_i})$. It remains to show that the columns of E_i and E_j, $i \neq j$, are orthogonal in the sense that all ordered pairs of symbols occur equally often. Without loss of generality, consider E_1 and E_2. Let $\mathbf{a}_1 = (a_{11}, \ldots, a_{1N})'$ and $\mathbf{a}_2 = (a_{21}, \ldots, a_{2N})'$ be columns from C_1 and C_2, respectively, and $\mathbf{d}_1 = (d_{11}, \ldots, d_{1M})'$ and $\mathbf{d}_2 = (d_{21}, \ldots, d_{2M})'$ be columns from $D(M, k_1; m_1)$ and $D(M, k_2; m_2)$, respectively. Then typical columns of E_1 and E_2 are given by $\mathbf{e}_1 = \mathbf{a}_1 \odot \mathbf{d}_1$ and $\mathbf{e}_2 = \mathbf{a}_2 \odot \mathbf{d}_2$, respectively. The ordered pairs arising from $\mathbf{e}_1$ and $\mathbf{e}_2$ are $(a_{1i} + d_{1j}, a_{2i} + d_{2j})$, $1 \le i \le N$, $1 \le j \le M$. Since $\mathbf{a}_1$ and $\mathbf{a}_2$ are orthogonal, it is clear that for any fixed j, all possible ordered pairs of symbols arise equally often among $\{a_{1i} + d_{1j}, a_{2i} + d_{2j} : 1 \le i \le N\}$. This shows the orthogonality of $\mathbf{e}_1$ and $\mathbf{e}_2$ and completes the proof. ■

Starting from this basic result, Wang and Wu (1991) proposed a general method of construction of asymmetric orthogonal arrays, consisting of the following steps:

Step 1. Construct the orthogonal array $OA(MN, \sum_{i=1}^{u} k_i n_i, {m_1}^{k_1 n_1} \times {m_2}^{k_2 n_2} \times \cdots \times {m_u}^{k_u n_u}, 2)$, as in Theorem 4.4.1. Call this array D_1.

Step 2. Let L^* denote an orthogonal array $OA(M, \sum_{j=1}^{p} r_j, {q_1}^{r_1} \times {q_2}^{r_2} \times \cdots \times {q_p}^{r_p}, 2)$ and $L = \mathbf{0}_N \odot L^* = [L^{*\prime} \vdots \cdots \vdots L^{*\prime}]'$, where L^* is repeated N times. Then $[D_1 \vdots L]$ is an orthogonal array $OA(MN, \sum k_i n_i + \sum r_j, \prod_{i=1}^{u} {m_i}^{k_i n_i} \times \prod_{j=1}^{p} {q_j}^{r_j}, 2)$.

As noted in Wang and Wu (1991), Steps 1 and 2 yield a tight orthogonal array provided that (a) the orthogonal arrays A of Theorem 4.4.1 and L^* of Step 2 are both tight, and (b) $k_i = M$ for each i (see (3.5.2)). Using the above method, they constructed several families of asymmetric orthogonal arrays. These are discussed below. In what follows, a difference matrix $D(N, N; 2)$, arising from a Hadamard matrix H_N (see Section 3.5), will be denoted by Δ_N and, without loss of generality, the initial column of Δ_N will be assumed to consist only of zeros. Then as in Theorem 3.2.2, the $N \times (N-1)$ array ${\Delta_N}^*$, obtained by deleting the initial column of Δ_N, represents a symmetric orthogonal array $L_N(2^{N-1})$.

Theorem 4.4.2. *Let N and T be Hadamard numbers and suppose $s < min(N, T)$. The existence of an orthogonal array $OA(N, s + \sum r_j, 2^s \times q_1^{r_1} \times q_2^{r_2} \times \cdots \times q_m^{r_m}, 2)$, say G, with the first s columns coinciding with those of $\Delta_N{}^*$, implies the existence of the array $OA(NT, N(T-1) - s + \sum r_j, 4^s \times 2^{N(T-1)-2s} \times \prod_{j=1}^m q_j{}^{r_j}, 2)$.*

Proof. Since T and N are Hadamard numbers, we have the array $OA(NT, N(T-1) + s + \sum r_j, 2^{N(T-1)+s} \times \prod_{j=1}^m q_j{}^{r_j}, 2)$, constructed as

$$E = [\Delta_T{}^* \odot \Delta_N \,\vdots\, \mathbf{0}_T \odot G].$$

Now let

$$\Delta_T{}^* = [\mathbf{a}_1 \,\vdots\, \mathbf{a}_2 \,\vdots \ldots \vdots\, \mathbf{a}_s \,\vdots\, A],$$

$$\Delta_N{}^* = [\mathbf{b}_1 \,\vdots\, \mathbf{b}_2 \,\vdots \ldots \vdots\, \mathbf{b}_s \,\vdots\, B],$$

$$G = [\mathbf{b}_1 \,\vdots\, \mathbf{b}_2 \,\vdots \ldots \vdots\, \mathbf{b}_s \,\vdots\, C].$$

Then E contains the disjoint sets of columns $\{\mathbf{a}_i \odot \mathbf{0}_N, \mathbf{a}_i \odot \mathbf{b}_i, \mathbf{0}_T \odot \mathbf{b}_i\}$, $1 \le i \le s$. Since

$$\mathbf{a}_i \odot \mathbf{0}_N + \mathbf{a}_i \odot \mathbf{b}_i \equiv \mathbf{0}_T \odot \mathbf{b}_i, \quad \text{mod } 2, \quad 1 \le i \le s,$$

exactly as in the last section, each such set of columns can be replaced by a four-symbol column. This yields the required array. ■

Remark 4.4.1. In Theorem 4.4.2, with $N \ge T, G = \Delta_N{}^*, s = T-1$ (so that $q_1 = \cdots = q_m = 2, \sum r_j = N - T$), we get the array $OA(NT, NT - 2T + 1, 4^{T-1} \times 2^{NT-3T+2}, 2)$ having the same parameters as the one constructed in Theorem 4.3.1. Also, with $N = 4u$, $T = 2$ and $s = 1$ in Theorem 4.4.2, we have an orthogonal array $OA(8u, 4u - 1 + \sum r_j, 4 \times 2^{4u-2} \times q_1{}^{r_1} \times \cdots \times q_m{}^{r_m}, 2)$ provided that an $OA(4u, 1 + \sum r_j, 2 \times q_1{}^{r_1} \times \cdots \times q_m{}^{r_m}, 2)$ exists; note that by a proper arrangement of the rows, the first column of the latter array always coincides with that of $\Delta_N{}^*$. In particular, using the array $OA(4u, 3, 2^2 \times (2u), 2)$, given by

$$[(0, 1)' \odot D(2u, 2; 2), \mathbf{0}_2 \odot (0, 1, \ldots, 2u - 1)'],$$

one gets the array $OA(8u, 4u + 1, 4 \times 2^{4u-1} \times (2u), 2)$, whose parameters are the same as those of the array constructed in Theorem 4.3.4. ■

Theorem 4.4.3. *Suppose that N and $T/2$ are Hadamard numbers, and let $s = min(N-1, T-1)$. Then the existence of an orthogonal array $OA(N, \sum r_j, {q_1}^{r_1} \times \cdots \times {q_m}^{r_m}, 2)$, say G, implies the existence of an $OA(NT, N(T-1) - 2s + \sum r_j, 4^s \times 2^{N(T-1)-3s} \times {q_1}^{r_1} \times \cdots \times {q_m}^{r_m}, 2)$.*

Proof. Let $R = T/2$. Since R is a Hadamard number, the array

$$A = [{\Delta_2}^* \odot \Delta_R \vdots \mathbf{0}_2 \odot {\Delta_R}^*]$$

represents an $OA(T, T-1, 2, 2)$. Writing $A = [\mathbf{a}_1 \vdots \mathbf{a}_2 \vdots \cdots \vdots \mathbf{a}_{T-1}]$, from the construction it follows that

$$\mathbf{a}_1 + \mathbf{a}_{i+1} \equiv \mathbf{a}_{R+i} \pmod 2, \qquad 1 \le i \le R-1. \tag{4.4.1}$$

Since N is a Hadamard number, the construction

$$E = [A \odot \Delta_N \vdots \mathbf{0}_T \odot G] \tag{4.4.2}$$

yields an $OA(NT, N(T-1) + \sum r_j, 2^{N(T-1)} \times \prod_{j=1}^m {q_j}^{r_j}, 2)$. Let

$$\Delta_N = [\mathbf{0}_N \vdots \mathbf{b}_1 \vdots \cdots \vdots \mathbf{b}_s \vdots B],$$

where B is $N \times (N-s-1)$. From (4.4.1), observe that

$$\mathbf{a}_1 \odot \mathbf{0}_N + \mathbf{a}_2 \odot \mathbf{b}_1 \equiv \mathbf{a}_{R+1} \odot \mathbf{b}_1 \bmod 2,$$
$$\mathbf{a}_1 \odot \mathbf{b}_{i+1} + \mathbf{a}_{i+1} \odot \mathbf{0}_N \equiv \mathbf{a}_{R+i} \odot \mathbf{b}_{i+1} \bmod 2,$$
$$1 \le i \le min(s-1, R-1), \text{ if } s > 1,$$

and

$$\mathbf{a}_1 \odot \mathbf{b}_{R+i} + \mathbf{a}_{i+1} \odot \mathbf{b}_{R+i} \equiv \mathbf{a}_{R+i} \odot \mathbf{0}_N \quad \bmod 2,$$
$$1 \le i \le s-R, \quad \text{if } s > R.$$

From these relations, it follows that one can find s disjoint triplets of two-symbol columns in (4.4.2) which can, as before, be replaced by s four-symbol columns. This leads to the required orthogonal array. ∎

Remark 4.4.2. If we take $m = r_1 = 1$ and $q_1 = N$ in Theorem 4.4.3, then we get the array $OA(NT, N(T-1) - 2s + 1, N \times 4^s \times 2^{N(T-1)-3s}, 2)$ considered in Theorem 4.3.2. Thus Theorem 4.4.3 leads to a proof of Theorem 4.3.2. ∎

Theorem 4.4.4. *Let N and $T/2$ be Hadamard numbers and $s = min(N - 1, T - 1)$. Then the existence of $OA(N, 1 + \sum r_j, 2 \times {q_1}^{r_1} \times \cdots \times {q_m}^{r_m}, 2)$ implies the existence of an $OA(NT, N(T-1) - 2h - 5 + \sum r_j, 8 \times 4^h \times 2^{N(T-1)-3h-6} \times \prod_{j=1}^{m} {q_j}^{r_j}, 2)$, where $h = \max(0, s-3)$.*

Proof. We proceed as in the proof of Theorem 4.4.3, using the same notation. Analogous to (4.4.2), we can construct $OA(NT, N(T-1) + 1 + \sum r_j, 2^{N(T-1)+1} \times \prod_{j=1}^{m} {q_j}^{r_j}, 2)$. Rearrange the rows of $OA(N, 1 + \sum r_j, 2 \times \prod_{j=1}^{m} {q_j}^{r_j}, 2)$, so that the first two-symbol column coincides with $\mathbf{b}_1$. Then by (4.4.1), the seven columns

$$[[\mathbf{a}_1, \mathbf{a}_2, \mathbf{a}_{R+1}] \odot [\mathbf{0}_N, \mathbf{b}_1], \mathbf{0}_T \odot \mathbf{b}_1],$$

together with a column of all zeros, form a group under addition modulo 2. Hence along the line of Remark 4.3.3, these seven columns can be replaced by an 8-symbol column.

Provided $s > 3$, we are left with $s - 3$ triplets of two-symbol columns according to the construction in Theorem 4.4.3. Replacing each such triplet by a four-symbol column, we get the required array. ■

Several other families of asymmetric orthogonal arrays can be obtained from these general classes of arrays by the procedures of collapsing and replacement. Since these are quite straightforward, we do not elaborate on these aspects. For a modification of the procedure of Wang and Wu (1991), giving in particular an $OA(4m^2, 2m+1, (2m)^3 \times m^{2m-2}, 2)$ whenever m is an odd prime or prime power, we refer to Mandeli (1995). The work of Wang (1996) is also of related interest.

4.5 USE OF RESOLVABLE ARRAYS

The use of *resolvable* orthogonal arrays for the construction of asymmetric orthogonal arrays was first made by Chacko and Dey (1981). Subsequently the resolvability of orthogonal arrays was exploited by Gupta, Nigam and Dey (1982), Dey and Agrawal (1985) and Hedayat, Pu, and Stufken (1992) for the construction of asymmetric orthogonal arrays of strength two. For a general possibly asymmetric orthogonal array, the definition of resolvability is as follows (see Hedayat et al. 1992):

Definition 4.5.1. *An orthogonal array $OA(N, \sum n_i, {m_1}^{n_1} \times {m_2}^{n_2} \times \cdots \times {m_t}^{n_t}, 2)$ is said to be $(\beta_1 \times \beta_2 \times \cdots \times \beta_t)$-resolvable if its rows can be partitioned into $N/\beta_1 m_1$ sets of $\beta_1 m_1$ rows each such that for any m_i-symbol* *column, each possible symbol occurs β_i times within the rows of every set $(i = 1, 2, \ldots, t)$.*

Clearly, for each $i = 1, 2, \ldots, t$, $\beta_i m_i$ does not depend on i and $\beta_i m_i$ divides N. If $m_1 = m_2 = \cdots = m_t = m$ and $\beta_1 = \beta_2 = \cdots = \beta_t = \beta$, then a $(\beta_1 \times \beta_2 \times \cdots \times \beta_t)$-resolvable array reduces to a (symmetric) β-resolvable array.

As an example, consider the following resolvable array $OA(16, 8, 4^2 \times 2^6, 2)$.

Example 4.5.1

$$OA(16, 8, 4^2 \times 2^6, 2) = \left[\begin{array}{cccccccc}
0 & 2 & 0 & 1 & 0 & 1 & 1 & 1 \\
3 & 1 & 0 & 0 & 1 & 1 & 0 & 0 \\
2 & 0 & 1 & 1 & 1 & 0 & 0 & 1 \\
1 & 3 & 1 & 0 & 0 & 0 & 1 & 0 \\
\hline
3 & 0 & 0 & 1 & 0 & 0 & 1 & 0 \\
0 & 3 & 0 & 0 & 1 & 0 & 0 & 1 \\
1 & 2 & 1 & 1 & 1 & 1 & 0 & 0 \\
2 & 1 & 1 & 0 & 0 & 1 & 1 & 1 \\
\hline
0 & 0 & 1 & 0 & 0 & 1 & 0 & 0 \\
3 & 3 & 1 & 1 & 1 & 1 & 1 & 1 \\
1 & 1 & 0 & 1 & 0 & 0 & 0 & 1 \\
2 & 2 & 0 & 0 & 1 & 0 & 1 & 0 \\
\hline
0 & 1 & 1 & 1 & 1 & 0 & 1 & 0 \\
1 & 0 & 0 & 0 & 1 & 1 & 1 & 1 \\
3 & 2 & 1 & 0 & 0 & 0 & 0 & 1 \\
2 & 3 & 0 & 1 & 0 & 1 & 0 & 0
\end{array}\right]. \qquad \blacksquare$$

We now have the following result:

Theorem 4.5.1. *Let there exist a $\prod_{i=1}^{u} \beta_i$-resolvable orthogonal array A, $OA(N, \sum_{i=1}^{u} n_i, \prod_{i=1}^{u} m_i^{n_i}, 2)$, and another orthogonal array, B, $OA(N/\beta_1 m_1, \sum_{i=u+1}^{t} n_i, \prod_{i=u+1}^{t} m_i^{n_i}, 2)$. Then there exists an orthogonal array $OA(N, \sum_{i=1}^{t} n_i, \prod_{i=1}^{t} m_i^{n_i}, 2)$.*

Proof. As in Definition 4.5.1, partition the rows of the resolvable array A into $N/\beta_1 m_1$ sets of $\beta_1 m_1$ rows each. For $1 \le i \le N/\beta_1 m_1$, augment each row in the ith set by the ith row of B to get the required orthogonal array. ■

For instance, with A as in Example 4.5.1 and B chosen as a symmetric $OA(4, 3, 2, 2)$, Theorem 4.5.1 yields a tight $OA(16, 11, 4^2 \times 2^9, 2)$. In particular, the trivial choice of B given by a single column of $N/\beta_1 m_1$ distinct symbols leads to the following corollary (see Dey and Agrawal, 1985).

Corollary 4.5.1. *The existence of a $\prod_{i=1}^{u}\beta_i$-resolvable orthogonal array $OA(N, \sum_{i=1}^{u} n_i, \prod_{i=1}^{u} m_i^{n_i}, 2)$ implies that of an orthogonal array $OA(N, \sum n_i + 1, \prod_{i=1}^{u} m_i^{n_i} \times (N/\beta_1 m_1), 2)$.* ■

The constructive proof of Theorem 4.5.1 resembles the technique of replacement introduced in Section 4.2. In fact, as one can easily observe, deletion of a column in an orthogonal array of strength two yields a resolvable array (with sets of rows corresponding to the symbols of the deleted column), and as such, the technique of replacement essentially represents an application of Theorem 4.5.1. It is also not difficult to see that orthogonal arrays given by Theorem 4.4.1 are resolvable. Hence, Steps 1 and 2 of Wang and Wu's (1991) construction, as discussed in the last section, can be justified via Theorem 4.5.1.

In general, difference matrices are very useful in the construction of resolvable symmetric orthogonal arrays. If Δ represents a difference matrix $D(\lambda m, n; m)$ and $\Delta_0, \Delta_1, \ldots, \Delta_{m-1}$ are as in the proof of Theorem 3.5.1, then developing Δ as $[\Delta_0', \Delta_1', \ldots, \Delta'_{m-1}]'$, we have a 1-resolvable symmetric array $OA(\lambda m^2, n, m, 2)$. By Theorem 4.5.1, this can be used to generate asymmetric orthogonal arrays. In particular, following Corollary 4.5.1, or Steps 1 and 2 of Wang and Wu's (1991) construction,

$$\begin{bmatrix} \Delta_0' & \Delta_1' & \cdots & \Delta'_{m-1} \\ \mathbf{a}' & \mathbf{a}' & \cdots & \mathbf{a}' \end{bmatrix}',$$

where $\mathbf{a} = (1, 2, \ldots, \lambda m)'$, represents an array $OA(\lambda m^2, n+1, m^n \times (\lambda m), 2)$. Thus we have

Lemma 4.5.1. *The existence of a difference matrix $D(\lambda m, n; m)$ implies that of an orthogonal array $OA(\lambda m^2, n+1, m^n \times (\lambda m), 2)$.* ■

Example 4.5.2. By Lemma 4.5.1, the difference matrix $D(6, 6; 3)$, shown in (3.5.1), leads to a tight array $OA(18, 7, 3^6 \times 6, 2)$ which is shown below:

$$OA(18, 7, 3^6 \times 6, 2) = \begin{bmatrix} 022011 & 100122 & 211200 \\ 121020 & 202101 & 010212 \\ 112002 & 220110 & 001221 \\ 000000 & 111111 & 222222 \\ 201012 & 012120 & 120201 \\ 210021 & 021102 & 102210 \\ 123456 & 123456 & 123456 \end{bmatrix}'.$$ ■

Lemma 4.5.1, when used in conjunction with Lemmas 3.5.1, 3.5.2, and Remark 3.5.3, yields the following result giving tight asymmetric orthogonal arrays:

Theorem 4.5.2.

(*a*) *(Gupta, Nigam, and Dey, 1982). If* $\lambda(\geq 2)$ *and* $m(\geq 2)$ *are powers of the same prime, then there exists an array* $OA(\lambda m^2, \lambda m+1, m^{\lambda m} \times (\lambda m), 2)$.

(*b*) *If* $m(\geq 2)$ *is a prime or prime power, then there exist arrays*

(*i*) $OA(2m^2, 2m+1, m^{2m} \times (2m), 2)$,

(*ii*) $OA(4m^2, 4m+1, m^{4m} \times (4m), 2)$. ■

Next consider the array $OA(2m^r, q, m, 2)$, where m is a prime or prime power, $r(\geq 2)$ is a positive integer and $q = 2(m^r - 1)/(m-1) - 1$, constructed by Addelman and Kempthorne (1961b) (*vide* Remark 3.5.4). This array has the property that the $2m^r$ rows can be partitioned into $2m$ sets of rows, each set containing m^{r-1} rows, such that within each set, one column is made up of only one of the m symbols while the other columns within a set contain each of the m symbols equally often. Omitting the column that contains the same symbol within a set, we get a resolvable array. Hence by Corollary 4.5.1, we have the following result due to Chacko and Dey (1981):

Theorem 4.5.3. *If* $m(\geq 2)$ *is a prime or prime power then, for positive integral* $r(\geq 2)$, *the tight array* $OA(2m^r, q, m^{q-1} \times (2m), 2)$ *exists, where* $q = [2(m^r - 1)/(m - 1)] - 1$. ■

Suen (1989b) discussed a method of constructing resolvable two-symbol symmetric orthogonal arrays of strength two. In conjunction with Corollary 4.5.1, his findings can also yield asymmetric orthogonal arrays.

Hedayat et al. (1992) constructed some general families of asymmetric orthogonal arrays by repeated application of the following result which is similar to Theorem 4.5.1 and has an identical proof (see Bose and Bush, 1952).

Theorem 4.5.4. *The existence of a* $\prod_{i=1}^{u} \beta_i$*-resolvable array* $OA(N, \sum_{i=1}^{u} n_i, \prod_{i=1}^{u} m_i^{n_i}, 2)$ *and a* $\prod_{i=u+1}^{t} \beta_i$*-resolvable array* $OA(N/\beta_1 m_1, \sum_{i=u+1}^{t} n_i, \prod_{i=u+1}^{t} m_i^{n_i}, 2)$ *implies the existence of a* $(\prod_{i=1}^{u} \beta_i \beta_t m_t \times \prod_{i=u+1}^{t} \beta_i \beta_1 m_1)$*-resolvable array* $OA(N, \sum_{i=1}^{t} n_i, \prod_{i=1}^{t} m_i^{n_i}, 2)$. ■

In order to facilitate the presentation of results due to Hedayat et al. (1992), we introduce some notation. Let $m(\geq 2)$ be a prime or prime power, n be a positive integer and $v_0, v_1, \ldots, v_r$ be positive integers such that $1 = v_0 < v_1 < \cdots < v_r \le n$ and $v_{i-1} \mid v_i$ $(1 \le i \le r)$. Then there exist unique nonnegative integers $c_0, c_1, \ldots, c_r$ satisfying

$$n = \sum_{i=0}^{r} c_i v_i, \quad 0 \le \sum_{i=0}^{j-1} c_i v_i < v_j, \qquad j = 1, 2, \ldots, r.$$

Define nonnegative integers $b_0, b_1, \ldots, b_r$ as

$$b_0 = 0,$$
$$b_i = c_{i-1} v_{i-1} + b_{i-1} \qquad (i = 1, 2, \ldots, r),$$

and let

$$u_{r+1} = n,$$
$$u_i = v_i + b_i, \qquad (i = 0, 1, \ldots, r),$$
$$n_i = \frac{(m^{u_{i+1}} - m^{u_i})}{(m^{v_i} - 1)}, \quad f_i = 2n_i \qquad (i = 0, 1, \ldots, r).$$

Furthermore define

$$n_0{}^* = n_0 - \frac{(m^{v_1} - m^{v_0})}{(m^{v_0} - 1)},$$

$$n_1{}^* = \begin{cases} n_1 + 1 & \text{if } c_1 = 0 \text{ and } b_1 > 0, \\ n_1 - \dfrac{(m^{v_2} - m^{v_1})}{(m^{v_1} - 1)} & \text{otherwise,} \end{cases}$$

$$n_i{}^* = \begin{cases} n_i + 1 & \text{if } c_{i-1} > 0 \text{ or } b_{i-1} = 0 \text{ and } c_i = 0, b_i > 0, \\ n_i & \text{if } c_{i-1} = 0, b_{i-1} > 0 \text{ and } c_i = 0, \\ n_i - \dfrac{(m^{v_{i+1}} - m^{v_i})}{(m^{v_i} - 1)} & \text{if } c_{i-1} > 0 \text{ or } b_{i-1} = 0 \text{ and } c_i > 0 \text{ or } b_i = 0, \\ n_i - \dfrac{(m^{v_{i+1}} - m^{v_i})}{(m^{v_i} - 1)} - 1 & \text{if } c_{i-1} = 0, b_{i-1} > 0 \text{ and } c_i > 0, \end{cases}$$

where $i = 2, 3, \ldots, r - 1$,

$$n_r{}^* = \begin{cases} n_r + 1 & \text{if } c_{r-1} > 0 \text{ or } b_{r-1} = 0, \\ n_r & \text{if } c_{r-1} = 0 \text{ and } b_{r-1} > 0. \end{cases}$$

Similarly let

$$f_0{}^* = \begin{cases} f_0 + 1 & \text{if } c_0 = 0, \\ f_0 - \dfrac{(m^{v_1} - m^{v_0})}{(m^{v_0} - 1)} & \text{if } c_0 > 0, \end{cases}$$

$$f_i^* = \begin{cases} f_i + 1 & \text{if } c_{i-1} > 0, \quad c_i = 0, \\ f_i & \text{if } c_{i-1} = 0, \quad c_i = 0, \\ f_i - \dfrac{(m^{v_{i+1}} - m^{v_i})}{(m^{v_i} - 1)} & \text{if } c_{i-1} > 0, \quad c_i > 0, \\ f_i - \dfrac{(m^{v_{i+1}} - m^{v_i})}{(m^{v_i} - 1)} - 1 & \text{if } c_{i-1} = 0, \quad c_i > 0, \end{cases}$$

where $i = 1, 2, \ldots, r - 1$,

$$f_r^* = \begin{cases} f_r + 1 & \text{if } c_{r-1} > 0, \\ f_r & \text{if } c_{r-1} = 0. \end{cases}$$

We are now in a position to present the following results due to Hedayat et al. (1992).

Theorem 4.5.5. *Let $m, n, v_0, v_1, \ldots, v_r, n_0^*, n_1^*, \ldots, n_r^*$ be as stated above. Then an orthogonal array $OA(m^n, \sum_{i=0}^{r} l_i, (m^{v_0})^{l_0} \times (m^{v_1})^{l_1} \times \cdots \times (m^{v_r})^{l_r}, 2)$ exists if the following inequalities hold:*

$$\sum_{i=j}^{r} l_i(m^{v_i} - 1) \leq \sum_{i=j}^{r} n_i^*(m^{v_i} - 1), \qquad j = 0, 1, \ldots, r. \qquad \blacksquare$$

Theorem 4.5.6. *Let $m, n, v_0, v_1, \ldots, v_r, f_0^*, f_1^*, \ldots, f_r^*$ be as stated above. Then an array $OA(2m^n, \sum_{i=0}^{r} l_i, (m^{v_0})^{l_0} \times (m^{v_1})^{l_1} \times \cdots \times (m^{v_r})^{l_r}, 2)$ exists if the following inequalities hold:*

$$\sum_{i=j}^{r} l_i(m^{v_i} - 1) \leq \sum_{i=j}^{r} f_i^*(m^{v_i} - 1), \qquad j = 0, 1, \ldots, r. \qquad \blacksquare$$

The inequalities in Theorem 4.5.5 hold trivially if $l_i = n_i^* (0 \leq i \leq r)$ in which case one gets a tight orthogonal array. Similarly the inequalities in Theorem 4.5.6 are satisfied if $l_i = f_i^* \quad (0 \leq i \leq r)$. We refer to the original paper for constructive proofs of these two results.

Example 4.5.3. To illustrate an application of Theorem 4.5.5, let $m = 2$, $n = 9, r = 3, v_0 = 1, v_1 = 2, v_2 = 4, v_3 = 8$. Then

$$c_0 = 1, c_1 = 0, c_2 = 0, c_3 = 1,$$

$$b_0 = 0, b_1 = 1, b_2 = 1, b_3 = 1,$$

$$u_0 = 1, u_1 = 3, u_2 = 5, u_3 = 9, u_4 = 9,$$
$$n_0 = 6, n_1 = 8, n_2 = 32, n_3 = 0,$$
$$n_0{}^* = 4, n_1{}^* = 9, n_2{}^* = 32, n_3{}^* = 0.$$

Hence by Theorem 4.5.5, an array $OA(512, l_0 + l_1 + l_2, 2^{l_0} \times 4^{l_1} \times 16^{l_2}, 2)$ can be constructed provided that

$$l_0 + 3l_1 + 15l_2 \leq 511, \quad l_1 + 5l_2 \leq 169, \qquad l_2 \leq 32.$$

In particular, if $l_i = n_i{}^*$ for each i, then these inequalities are satisfied, and we get a tight array $OA(512, 45, 2^4 \times 4^9 \times 16^{32}, 2)$. For the details of actual construction, we again refer to Hedayat et al. (1992). ■

Before concluding this section, we state the following result, resembling Theorem 4.4.1, due to Dey and Midha (1996). For a proof of this result, which also exploits the notion of resolvability, the reader is referred to the original source.

Theorem 4.5.7. *The existence of an orthogonal array* $OA(N, \sum_{i=1}^{u} n_i, m_1{}^{n_1} \times \cdots \times m_u{}^{n_u}, 2)$ *and difference matrices* $D(M, k_i; m_i)$, $i = 1, 2, \ldots, u$ *implies the existence of an orthogonal array* $OA(MN, 1 + \sum_{i=1}^{u} k_i n_i - k_1, (Mm_1) \times m_1{}^{k_1(n_1-1)} \times m_2{}^{k_2 n_2} \times \cdots \times m_u{}^{k_u n_u}, 2)$. ■

4.6 MORE ON THE METHOD OF GROUPING

The method of grouping for the construction of asymmetric orthogonal arrays has been informally introduced in Sections 4.3 and 4.4. As discussed there, if a two-symbol symmetric orthogonal array with symbols -1 and $+1$ (or, 0 and 1) contains a triplet of columns, the Hadamard product (or, sum reduced modulo 2) of any two of which equals the third, then such a triplet can be replaced by a four-symbol column. Furthermore, if the two-symbol array contains p such disjoint triplets of columns, then these can be replaced by as many four-symbol columns without affecting orthogonality.

Addelman (1962a) employed the above technique for constructing asymmetric orthogonal arrays of the type $OA(N, n_1 + n_2, 2^{n_1} \times 4^{n_2}, 2)$, where $n_1 + 3n_2 = N - 1$, starting from symmetric arrays $OA(N, N - 1, 2, 2)$. For $N = 2^r$ $(r \geq 2)$, he also showed that n_2, the number of 4-symbol columns, satisfies

$$n_2 \leq \begin{cases} \dfrac{N-1}{3} & \text{if } r \text{ is even,} \\ \dfrac{N-5}{3} & \text{if } r \text{ is odd.} \end{cases} \tag{4.6.1}$$

Wu (1989) gave a general procedure for grouping the columns and also showed how the upper bounds (4.6.1) can actually be achieved; see Theorem 4.6.2 below.

Wu's (1989) findings were strengthened and generalized by Wu et al. (1992) who introduced a grouping scheme with reference to the symmetric orthogonal array $OA(m^r, (m^r - 1)/(m - 1), m, 2)$, say A, constructed in Theorem 3.4.3. Recall that here m is a prime or prime power and that the symbols in A are elements of $GF(m)$. Furthermore, since A is an orthogonal array of strength two, no column of A equals the null vector, and no two distinct columns of A are proportional to each other. For $k \leq r$, consider a set S of $\mu_k = (m^k - 1)/(m - 1)$ distinct columns of A such that S is closed, up to proportionality, under the formation of non-null linear combinations of the columns therein. Then there are k linearly independent columns (in S) that span S. Hence it can be seen that the $m^r \times \mu_k$ subarray of A given by the columns in S has m^k distinct rows each repeated m^{r-k} times and that these rows can be replaced by m^k distinct symbols (i.e., the μ_k columns in S can be replaced by a single m^k-symbol column) retaining orthogonality (see Addelman, 1962a). If there are p disjoint sets of columns each having the above property, then each such set of columns can be replaced by an m^k-symbol column without affecting orthogonality. Wu et al. (1992) developed a systematic method for identifying disjoint sets of columns as indicated above and hence obtained the following result:

Theorem 4.6.1. *Let $m(\geq 2)$ be a prime or prime power and k and $r(\geq 2)$ be positive integers such that $r = kq + p$, $q \geq 1$, $0 \leq p \leq k - 1$. Then tight asymmetric orthogonal arrays of the type $OA(m^r, n_1 + n_2, m^{n_1} \times (m^k)^{n_2}, 2)$ exist for any integers n_1 and n_2 satisfying*

$$n_1(m - 1) + n_2(m^k - 1) = m^r - 1,$$

and

$$n_2 = 1, 2, \ldots, \frac{(m^r - m^{k+p})}{(m^k - 1)} + 1. \qquad \blacksquare$$

Example 4.6.1. Let $m = r = 3$ and $k = 2$. Then, following the proof of Theorem 3.4.3, the matrix C considered there is given by

$$C = \begin{bmatrix} 111 & 111 & 111 & 0000 \\ 000 & 111 & 222 & 1110 \\ 012 & 012 & 012 & 0121 \end{bmatrix}.$$

As in Lemma 3.4.1, consideration of all linear combinations of the rows of C yields a symmetric $OA(27, 13, 3, 2)$, say A. With $k = 2$, we have $\mu_k = 4$. Observe that the first, fourth, seventh and tenth columns of C, and hence those of A, form a set which is closed, up to proportionality, under the formation of non-null linear combinations. Hence replacing these four columns of A by a single $9(= 3^2)$-symbol column, one gets an $OA(27, 10, 9 \times 3^9, 2)$, which is tight. ■

With $m = k = 2$, Theorem 4.6.1 yields the following result, due to Wu (1989), showing the attainabilty of the bounds in (4.6.1):

Theorem 4.6.2. *Let $r(\geq 2)$ be a positive integer. Then tight asymmetric orthogonal arrays of the type $OA(2^r, n_1 + n_2, 2^{n_1} \times 4^{n_2}, 2)$ exist for any integers n_1 and n_2 satisfying $n_1 + 3n_2 = 2^r - 1$ and $n_2 = 1, 2, \ldots, t$, where*

$$t = \begin{cases} \dfrac{(2^r - 1)}{3} & \text{if } r \text{ is even,} \\ \dfrac{(2^r - 5)}{3} & \text{if } r \text{ is odd.} \end{cases}$$ ■

We refer to Wu et al. (1992) for a constructive proof of Theorem 4.6.1. These authors also generalized Theorem 4.6.1 to obtain asymmetric orthogonal arrays of the type $OA(m^r, \sum_{i=0}^{u} n_i, m^{n_0} \times (m^{k_1})^{n_1} \times \cdots \times (m^{k_u})^{n_u}, 2)$, where m is a prime or prime power and the n_i's and k_i's are positive integers. We refer to Wu et al. (1992) and Hedayat et al. (1992) for a comparative discussion on the results reported therein.

4.7 ARRAYS OF HIGHER STRENGTH

In the earlier sections we discussed several methods of obtaining asymmetric orthogonal arrays of strength *two*. In this section we primarily consider arrays of strength three and observe, in particular, how Hadamard matrices can be useful in their construction. A brief remark will also be made on arrays of still higher strength. We begin by presenting the following result due to Dey (1993).

Theorem 4.7.1. *Let $N \geq 4$ and $T < N$ be Hadamard numbers. Then there exists an $OA(4NT, NT - T^2 + T, 4^T \times 2^{T(N-T)}, 3)$.*

Proof. Write H_N, a Hadamard matrix of order N, as

$$H_N = [\mathbf{1}_N \vdots \boldsymbol{a}_1 \vdots \boldsymbol{a}_2 \vdots \cdots \vdots \boldsymbol{a}_T \vdots B_T],$$

where B_T is $N \times (N - T - 1)$. Define

$$B^* = [B_T \vdots \mathbf{1}_N], \quad A_T = \begin{bmatrix} H_T \\ H_T \end{bmatrix}, \quad D_1 = \begin{bmatrix} H_T \\ -H_T \end{bmatrix}.$$

For $i = 1, \ldots, T$, replace the -1's in the ith column of D_1 by $\boldsymbol{a}_i$ and $+1$'s by $3\boldsymbol{a}_i$ to obtain a matrix D_2 of order $2NT \times T$. Define

$$D_0 = [D_2 \vdots A_T \otimes B^*],$$

and in the spirit of Remark 3.3.1, let

$$D = \begin{bmatrix} D_0 \\ -D_0 \end{bmatrix}.$$

Then it can be verified that D is the required orthogonal array. ■

Remark 4.7.1. The above construction is valid even for $T = 1$, in which case we get the array $OA(4N, N, 4 \times 2^{N-1}, 3)$ constructed earlier by Margolin (1969). For $T = 2, 4$, Theorem 4.7.1 yields the arrays $OA(8N, 2N - 2, 4^2 \times 2^{2N-4}, 3)$ and $OA(16N, 4N - 12, 4^4 \times 2^{4N-16}, 3)$, respectively. The first of these was obtained earlier by Agrawal and Dey (1983), while the second one is similar to one of their constructions (see Theorem 4.7.2 (a) below), but accommodates more four-symbol columns. From Theorem 2.6.2, it can be checked that Theorem 4.7.1 yields a tight array if and only if $T = 1$ or 2. ■

Theorem 4.7.2. *(Agrawal and Dey, 1983). Let $N(\geq 4)$ be a Hadamard number. Then there exist (a) an array $OA(16N, 4N - 9, 4^3 \times 2^{4N-12}, 3)$ and (b) a tight array $OA(32N, 4N - 6, 8^2 \times 2^{4N-8}, 3)$.*

Proof. Let $H_N = [\mathbf{1}_N \vdots B]$, where B is $N \times (N - 1)$. Permute the rows of B, if necessary, so that in the first column, the first $N/2$ elements are -1 and the rest are $+1$ and call the matrix so obtained B^*.

(a) Partition B^* as $B^* = [\boldsymbol{a}_1 \vdots \boldsymbol{a}_2 \vdots \boldsymbol{a}_3 \vdots B_4]$, where B_4 is $N \times (N-4)$. Let ${B_4}^* = [B_4 \vdots \mathbf{1}_N]$. Then the matrix

$$D_1 = \begin{bmatrix} {D_1}^* \\ -{D_1}^* \end{bmatrix},$$

where

$${D_1}^* = \begin{bmatrix} \boldsymbol{a}_1 & \boldsymbol{a}_2 & \boldsymbol{a}_3 & {B_4}^* & -{B_4}^* & -{B_4}^* & -{B_4}^* \\ 3\boldsymbol{a}_1 & \boldsymbol{a}_2 & \boldsymbol{a}_3 & {B_4}^* & {B_4}^* & -{B_4}^* & {B_4}^* \\ \boldsymbol{a}_1 & 3\boldsymbol{a}_2 & \boldsymbol{a}_3 & {B_4}^* & {B_4}^* & {B_4}^* & {B_4}^* \\ 3\boldsymbol{a}_1 & 3\boldsymbol{a}_2 & \boldsymbol{a}_3 & {B_4}^* & -{B_4}^* & {B_4}^* & -{B_4}^* \\ \boldsymbol{a}_1 & \boldsymbol{a}_2 & 3\boldsymbol{a}_3 & {B_4}^* & -{B_4}^* & {B_4}^* & {B_4}^* \\ 3\boldsymbol{a}_1 & \boldsymbol{a}_2 & 3\boldsymbol{a}_3 & {B_4}^* & {B_4}^* & {B_4}^* & -{B_4}^* \\ \boldsymbol{a}_1 & 3\boldsymbol{a}_2 & 3\boldsymbol{a}_3 & {B_4}^* & {B_4}^* & -{B_4}^* & -{B_4}^* \\ 3\boldsymbol{a}_1 & 3\boldsymbol{a}_2 & 3\boldsymbol{a}_3 & {B_4}^* & -{B_4}^* & -{B_4}^* & {B_4}^* \end{bmatrix},$$

can be seen to represent the array in (a).

(b) Now let $B_1 = [\boldsymbol{a}_3 \vdots B_4 \vdots \mathbf{1}_N]$ and consider the matrix

$$D_2 = \begin{bmatrix} {D_2}^* \\ -{D_2}^* \end{bmatrix},$$

where

$${D_2}^* = \begin{bmatrix} \boldsymbol{a}_1 & \boldsymbol{a}_2 & B_1 & -B_1 & B_1 & -B_1 \\ 3\boldsymbol{a}_1 & \boldsymbol{a}_2 & B_1 & B_1 & -B_1 & -B_1 \\ 5\boldsymbol{a}_1 & \boldsymbol{a}_2 & B_1 & -B_1 & -B_1 & B_1 \\ 7\boldsymbol{a}_1 & \boldsymbol{a}_2 & B_1 & B_1 & B_1 & B_1 \\ \boldsymbol{a}_1 & 3\boldsymbol{a}_2 & B_1 & B_1 & -B_1 & -B_1 \\ 3\boldsymbol{a}_1 & 3\boldsymbol{a}_2 & B_1 & -B_1 & B_1 & -B_1 \\ 5\boldsymbol{a}_1 & 3\boldsymbol{a}_2 & B_1 & B_1 & B_1 & B_1 \\ 7\boldsymbol{a}_1 & 3\boldsymbol{a}_2 & B_1 & -B_1 & -B_1 & B_1 \\ \boldsymbol{a}_1 & 5\boldsymbol{a}_2 & B_1 & -B_1 & -B_1 & B_1 \\ 3\boldsymbol{a}_1 & 5\boldsymbol{a}_2 & B_1 & B_1 & B_1 & B_1 \\ 5\boldsymbol{a}_1 & 5\boldsymbol{a}_2 & B_1 & -B_1 & B_1 & -B_1 \\ 7\boldsymbol{a}_1 & 5\boldsymbol{a}_2 & B_1 & B_1 & -B_1 & -B_1 \\ \boldsymbol{a}_1 & 7\boldsymbol{a}_2 & B_1 & B_1 & B_1 & B_1 \\ 3\boldsymbol{a}_1 & 7\boldsymbol{a}_2 & B_1 & -B_1 & -B_1 & B_1 \\ 5\boldsymbol{a}_1 & 7\boldsymbol{a}_2 & B_1 & B_1 & -B_1 & -B_1 \\ 7\boldsymbol{a}_1 & 7\boldsymbol{a}_2 & B_1 & -B_1 & B_1 & -B_1 \end{bmatrix}.$$

Then it can be seen that D_2 represents the array in (b). Its tightness is a simple consequence of Theorem 2.6.2. ■

Theorem 4.7.3. *(Gupta and Nigam, 1985). Let N and T be Hadamard numbers. Then there exists a tight array* $OA(NT^2, T(N-1)+2, T^2 \times 2^{T(N-1)}, 3)$.

Proof. Write $\mathbf{b}_1 = (1, 2, \ldots, T)'$, and let

$$\mathbf{b}_2 = (2, 3, \ldots, T, 1)', \quad \ldots, \quad \mathbf{b}_T = (T, 1, \ldots, T-2, T-1)'$$

be vectors obtained by cyclically permuting the elements of $\mathbf{b}_1$. Define the $NT^2 \times 1$ vector

$$\mathbf{b} = (\mathbf{b}_1{}' \otimes \mathbf{1}'_N, \mathbf{b}_2{}' \otimes \mathbf{1}'_N, \ldots, \mathbf{b}_T{}' \otimes \mathbf{1}'_N)'. \tag{4.7.1}$$

Let $H_N = [\mathbf{1}_N \vdots B]$ and define the $T \times 1$ vector $\mathbf{c} = (1, -1, 1, -1, \ldots, 1, -1)'$. As in Section 3.3, $\mathbf{c} \otimes H_T \otimes B$ gives a symmetric array $OA(NT^2, T(N-1), 2, 3)$. Hence it is not hard to see that

$$D = [\mathbf{b}_1 \otimes \mathbf{1}_T \otimes \mathbf{1}_N \vdots \mathbf{b} \vdots \mathbf{c} \otimes H_T \otimes B] \tag{4.7.2}$$

represents the required orthogonal array. Its tightness follows from Theorem 2.6.2. ■

Example 4.7.1. In Theorem 4.7.3, let $N = 2$, $T = 4$. Then with

$$\mathbf{b}_1 = (1, 2, 3, 4)', \quad \mathbf{b}_2 = (2, 3, 4, 1)', \quad \mathbf{b}_3 = (3, 4, 1, 2)', \quad \mathbf{b}_4 = (4, 1, 2, 3)',$$

$$B = \begin{bmatrix} 1 \\ -1 \end{bmatrix}, \quad H_T = \begin{bmatrix} 1 & 1 & 1 & 1 \\ 1 & -1 & 1 & -1 \\ 1 & 1 & -1 & -1 \\ 1 & -1 & -1 & 1 \end{bmatrix}, \quad \mathbf{c} = (1, -1, 1, -1)'$$

in (4.7.1) and (4.7.2), one gets a tight array $OA(32, 6, 4^2 \times 2^4, 3)$. An array with the same parameters can as well be obtained by taking $N = 4$, $T = 2$ in Theorem 4.7.1 (*vide* Remark 4.7.1). ■

The next result shows how, given an orthogonal array of strength three, further arrays can be generated.

Theorem 4.7.4. *Suppose that an orthogonal array A,* $OA(N, n, m_1 \times m_2 \times \cdots \times m_n, 3)$ *is available, and let T be a positive integer such that* $m_1 \mid T$. *Then there exists an array* $OA(NT/m_1, n, T \times m_2 \times \cdots \times m_n, 3)$.

Proof. Without loss of generality, we can write A as

$$A = \begin{bmatrix} \mathbf{a}_1' & \mathbf{a}_2' & \cdots & \mathbf{a}'_{m_1} \\ A_1' & A_2' & \cdots & A'_{m_1} \end{bmatrix}',$$

where, for $1 \leq i \leq m_1$, $\mathbf{a}_i$ is a vector with N/m_1 elements each equal to i. Clearly, then for each i, A_i represents an array $OA(N/m_1, n-1, m_2 \times \cdots \times m_n, 2)$. Hence, writing $u = N/m_1$, $q = T/m_1$, $\mathbf{a} = (1, 2, \ldots, T)'$ and $A^* = (A_1' \vdots A_2' \vdots \cdots \vdots A'_{m_1})'$, it is easy to see that

$$[\mathbf{a} \otimes \mathbf{1}_u \vdots \mathbf{1}_q \otimes A^*]$$

represents the required orthogonal array. ■

Remark 4.7.2. In Theorem 4.7.4, if $m_1 = \max_{1 \leq i \leq n} m_i$ and the initial array A is tight, then by Theorem 2.6.2, one can check that the resulting array is also tight. ■

Earlier, in Corollary 3.3.1, we noted, for every Hadamard number $N(\geq 4)$, the existence of a symmetric array $OA(2N, N, 2, 3)$ which is easily seen to be tight. Application of Theorem 4.7.4 to this array yields the following result due to Margolin (1969):

Corollary 4.7.1. *For every Hadamard number $N(\geq 4)$ and even integer $T(\geq 4)$, there exists a tight array $OA(NT, N, T \times 2^{N-1}, 3)$.* ■

Combining Theorem 4.7.4 with Theorem 3.4.1 (see also Remark 3.4.1), the following result is evident:

Corollary 4.7.2.

(a) *Let $m(\geq 2)$ be a prime or prime power. Then, for every positive integer T such that $m \mid T$, there exists an array $OA(m^2T, m+1, T \times m^m, 3)$.*

(b) *If in addition $m(\geq 4)$ is even, then for every positive integer T such that $m \mid T$, there exists a tight array $OA(m^2T, m+2, T \times m^{m+1}, 3)$.*

■

In fact numerous other applications of Theorem 4.7.4 can be imagined. For example, applying it to the tight array $OA(8N, 2N-2, 4^2 \times 2^{2N-4}, 3)$ discussed in Remark 4.7.1, one gets the tight array $OA(2NT, 2N-2, T \times$

$4 \times 2^{2N-4}, 3)$, where $4 \mid T$ and $N(\geq 4)$ is a Hadamard number. Similarly, for any Hadamard number $N(\geq 4)$, Theorem 4.7.4, in conjunction with Theorem 4.7.2, yields the arrays $OA(4NT_1, 4N - 9, T_1 \times 4^2 \times 2^{4N-12}, 3)$ and $OA(4NT_2, 4N - 6, T_2 \times 8 \times 2^{4N-8}, 3)$, where $4 \mid T_1$ and $8 \mid T_2$; the latter of these two arrays is tight. We refer to Dey (1985, ch. 4) for more discussion in this regard.

Example 4.7.2. To illustrate an application of Theorem 4.7.4, let A be the symmetric array $OA(8, 4, 2, 3)$ given by (see Remark 3.3.1)

$$A = [H_4' \vdots - H_4']' = \begin{bmatrix} \mathbf{1}_4' & -\mathbf{1}_4' \\ B' & -B' \end{bmatrix}',$$

where

$$B = \begin{bmatrix} 1 & -1 & 1 & -1 \\ 1 & 1 & -1 & -1 \\ 1 & -1 & -1 & 1 \end{bmatrix}'.$$

Let $T = 6$. Then $u = 4$, $q = 3$, $\mathbf{a} = (1, 2, 3, 4, 5, 6)'$, $A^* = [B', -B']'$, and following the proof of Theorem 4.7.4, a tight array $OA(24, 4, 6 \times 2^3, 3)$ is given by

$$\begin{bmatrix} \mathbf{b}_1' & \mathbf{b}_2' & \mathbf{b}_3' & \mathbf{b}_4' & \mathbf{b}_5' & \mathbf{b}_6' \\ B' & -B' & B' & -B' & B' & -B' \end{bmatrix}',$$

where for $i = 1, 2, \ldots, 6$, $\mathbf{b}_i = (i, i, i, i)'$. ■

With reference to Hadamard matrices of order $N = 2^r$, Addelman (1972) considered a method of grouping for the construction of arrays of the type $OA(N, n_1 + n_2, 2^{n_1} \times 4^{n_2}, 3)$. He, however, did not present a general construction procedure, and many of his examples are covered also by the results shown above. We refer to Addelman (1972) and Dey (1985) for details.

The literature on the construction of asymmetric orthogonal arrays of strength four or more is scanty. These arrays often involve a large number of rows and therefore may not be of much practical utility. Dey (1985, pp. 96–97) indicated a procedure, akin to Theorem 4.7.4, for the construction of asymmetric orthogonal arrays of strength four. The problem of finding asymmetric arrays of strength five or more in reasonable number of rows is largely unexplored.

EXERCISES

4.1. Verify the conclusion of Remark 4.3.3 for the case $\nu = 3$.

4.2. With reference to Section 4.4, verify that Steps 1 and 2 yield a tight orthogonal array provided that the orthogonal arrays A of Theorem 4.4.1 and L^* of Step 2 are both tight and that $k_i = M$ for each i.

4.3. Verify the result of Shrikhande (1964) stated in the proof of Theorem 4.4.1.

4.4. Follow Remark 4.4.1 to construct an orthogonal array $OA(24, 13, 4 \times 2^{11} \times 6, 2)$.

4.5. Use Theorem 2.6.2(a) and Lemma 4.5.1 to prove the inequality (3.5.2).

4.6. Prove Theorem 4.6.1 in the following special cases:

(a) $r \geq 3, k = 2, n_2 = 1$.

(b) $r \geq 4, k = 2, n_2 = 2$.

4.7. Verify Remark 4.7.2.

CHAPTER 5

Some Results on Nonexistence

5.1 INTRODUCTION

In Theorem 2.6.2 we had introduced Rao's bound. This bound is a fundamental necessary condition for the existence of an orthogonal array. Additional necessary conditions are implicit in the definition of such an array. Thus, for the existence of an $OA(N, n, m_1 \times \cdots \times m_n, 2)$, in view of Rao's bound, it is necessary that $N \geq 1 + \sum_{j=1}^{n}(m_j - 1)$. In addition N has to be an integral multiple of $m_i m_j$ for $1 \leq i < j \leq n$. However, even when these necessary conditions hold, there is no guarantee that an orthogonal array will exist. In this chapter we briefly review some nonexistence results on orthogonal arrays. Such results are important in the context of fractional factorial plans, since if in a given situation an orthogonal array is known to be nonexistent then the problem of finding an optimal fraction gets much more involved and calls for the development of appropriate additional techniques.

In Section 5.2 we consider an extended version of the Bose-Bush approach, which was proposed originally for symmetric orthogonal arrays. Some linear programming and enumerative bounds on orthogonal arrays are discussed in Section 5.3. Finally in Section 5.4 we deal with the existence problem for tight and nearly tight arrays.

5.2 BOSE-BUSH APPROACH

The following lemma is central to this approach.

Lemma 5.2.1. *Let $y_1, \ldots, y_u$ be integers with arithmetic mean $\bar{y}$ and ψ be the fractional part of $\bar{y}$. Then*

$$\frac{1}{u}\sum_{i=1}^{u}(y_i - \bar{y})^2 \geq \psi(1 - \psi).$$

Proof. Let y_0 be the largest integer that does not exceed $\bar{y}$. Then

$$\psi = \bar{y} - y_0. \tag{5.2.1}$$

Since $y_1, \ldots, y_u$ are all integers, we have $(y_i - y_0)(y_i - y_0 - 1) \geq 0$, namely

$$y_i^2 - (2y_0 + 1)y_i + y_0(y_0 + 1) \geq 0, \qquad 1 \leq i \leq u.$$

Summing the above over i, after a little algebra, we have

$$\frac{1}{u}\sum_{i=1}^{u}(y_i - \bar{y})^2 \geq (\bar{y} - y_0)(1 - \bar{y} + y_0) = \psi(1 - \psi), \tag{5.2.2}$$

by (5.2.1). ■

Consider now an orthogonal array $OA(N, n, m_1 \times \cdots \times m_n, 2)$. For $2 \leq i \leq N$, let h_i be the number of coincidences between the first and the ith rows of the array; namely h_i is the number of positions in which these two rows have the same symbol. Then

$$\sum_{i=2}^{N} h_i = \sum_{j=1}^{n}\left(\frac{N}{m_j} - 1\right), \quad \sum_{i=2}^{N}\binom{h_i}{2} = \sum\sum_{1 \leq j < j' \leq n}\left(\frac{N}{m_j m_{j'}} - 1\right). \tag{5.2.3}$$

The first identity in (5.2.3) is obvious. The second identity follows by counting in two ways the occurrences of ordered pairs of symbols arising from the first row in the corresponding positions of other rows. By (5.2.3),

$$\bar{h} = \frac{\sum_{i=2}^{N} h_i}{N - 1} = (N - 1)^{-1}\sum_{j=1}^{n}\left(\frac{N}{m_j} - 1\right), \tag{5.2.4}$$

and let ψ be the fractional part of $\bar{h}$. Then the following result holds:

Theorem 5.2.1. *For the existence of an* $OA(N, n, m_1 \times \cdots \times m_n, 2)$, *it is necessary that*

$$N(N - 1)^{-2}\left\{(N - 1)\sum_{j=1}^{n}\frac{m_j - 1}{m_j^2} - \left(\sum_{j=1}^{n}\left(\frac{m_j - 1}{m_j}\right)\right)^2\right\} \geq \psi(1 - \psi). \tag{5.2.5}$$

Proof. The left-hand side of (5.2.5) equals $(N - 1)^{-1}\sum_{i=2}^{N}(h_i - \bar{h})^2$, by (5.2.3). Since $h_2, \ldots, h_N$ are integers, the result now follows from Lemma 5.2.1. ■

Considering, in particular, symmetric orthogonal arrays, the following corollary is evident from (5.2.4) and Theorem 5.2.1:

Corollary 5.2.1. *For the existence of a symmetric orthogonal array $OA(N, n, m, 2)$, it is necessary that*

$$\frac{Nn(m-1)}{\{(N-1)m\}^2}\{N-1-n(m-1)\} \geq \psi(1-\psi), \tag{5.2.6}$$

where ψ is the fractional part of $n(N-m)/\{m(N-1)\}$. ■

Example 5.2.1. This example substantiates the comment made in Remark 3.5.2 that in an $OA(2m^2, n, m, 2)$ if $m \geq 3$, then $n \leq 2m+1$. If an $OA(2m^2, 2m+2, m, 2)$ exists, then the quantity ψ, as defined in Corollary 5.2.1, equals $2m/(2m^2-1)$, while the left-hand side of (5.2.6) equals $4(m^2-1)/((2m^2-1)^2)$. So (5.2.6) is violated whenever $m \geq 3$. This establishes the claim made in Remark 3.5.2. ■

Example 5.2.2. We employ Theorem 5.2.1 to prove the nonexistence of an $OA(36, 22, 2^9 \times 3^{13}, 2)$. For such an array, by (5.2.4), $\psi = 16/35$, whereas the left-hand side of (5.2.5) equals 234/1225. Hence (5.2.5) is violated and the nonexistence follows. ■

Using an approach similar to that employed in the context of Corollary 5.2.1, Bose and Bush (1952) obtained the following result.

Theorem 5.2.2. *Consider a symmetric $OA(\lambda m^2, n, m, 2)$. If $\lambda - 1$ is not an integral multiple of $m-1$, then for the existence of such an array, it is necessary that*

$$n \leq \left(\frac{\lambda m^2 - 1}{m-1}\right) - [\theta] - 1, \tag{5.2.7}$$

where

$$\theta = \frac{\sqrt{1+4m(m-1-b)} - (2m-2b-1)}{2}, \tag{5.2.8}$$

b is the remainder on dividing $\lambda - 1$ by $m-1$ and $[z]$ is the largest integer that does not exceed z. ■

Remark 5.2.1. The upper bound in (5.2.7) is called the Bose-Bush bound. A comparison between Corollary 5.2.1 and Theorem 5.2.2 is in order. For this purpose, we revisit the setup of Lemma 5.2.1 and note that analogously to (5.2.2),

$$\frac{1}{u}\sum_{i=1}^{u}(y_i - \bar{y})^2 \geq (\bar{y} - y)(1 - \bar{y} + y), \tag{5.2.9}$$

for every integer y. The choice $y = y_0$ that yields Lemma 5.2.1 and hence Corollary 5.2.1 corresponds to the sharpest bound of the form (5.2.9). Since Theorem 5.2.2 is also a consequence of (5.2.9), it follows that Corollary 5.2.1 is always at least as strong as Theorem 5.2.2. In fact, there can be situations, though infrequent, where Corollary 5.2.1 is stronger than Theorem 5.2.2. For example, the nonexistence of an $OA(2888, 157, 19, 2)$ follows from Corollary 5.2.1, but not from Theorem 5.2.2. ■

Lemma 5.2.2. *The existence of a symmetric $OA(\lambda m^g, n+1, m, g)$ implies that of an $OA(\lambda m^{g-1}, n, m, g-1)$.*

Proof. Given an $OA(\lambda m^g, n+1, m, g)$, one can write down the portion that has the same symbol in the first column and then delete the first column to get an $OA(\lambda m^{g-1}, n, m, g-1)$. ■

From Theorem 5.2.2 and Lemma 5.2.2, the following result, due to Bose and Bush (1952), is evident:

Theorem 5.2.3. *Consider a symmetric $OA(\lambda m^3, n, m, 3)$. If $\lambda - 1$ is not an integral multiple of $m - 1$, then for the existence of such an array it is necessary that $n \leq [(\lambda m^2 - 1)/(m-1)] - [\theta]$, where θ is as given by (5.2.8).* ■

Before concluding this section, we briefly indicate some other results. By Rao's bound, in an $OA(m^3, n, m, 3)$, $n \leq m+2$; in addition, if m is odd, then as noted in Remark 3.4.1, $n \leq m+1$. Hence by Lemma 5.2.2, it follows that if $g \geq 3$, then in an $OA(m^g, n, m, g)$, $n \leq m+g-1$ for even m and $n \leq m+g-2$ for odd m (*vide* Bush, 1952). As mentioned in Remark 3.4.1, Bush (1952) also showed that in an $OA(m^g, n, m, g)$, if $m \leq g$, then $n \leq g+1$ (see Mukerjee, 1979, for a simpler proof). Summarizing these results, we get the following theorem due to Bush (1952):

Theorem 5.2.4. *In an $OA(m^g, n, m, g)$ with $g \geq 3$,*

(*a*) $n \leq g+1$ if $m \leq g$;
(*b*) $n \leq m+g-1$ if $m > g$ and m is even;
(*c*) $n \leq m+g-2$ if $m > g$ and m is odd. ■

Hedayat and Stufken (1989) showed that the bounds in Theorem 5.2.4 are at least as sharp as Rao's bound. Kounias and Petros (1975) strengthened these bounds in special cases and, in particular, proved that (1) in an $OA(m^3, n, m, 3)$, if $m = 4u + 2, u \geq 1$ then $n \leq m$, and (2) in an $OA(m^4, n, m, 4)$, if $m(\geq 4)$ is even and $m \neq 0 \pmod{36}$ then $n \leq m + 1$.

The issue of existence of an $OA(2.3^g, n, 3, g)$, $g \geq 2$, has received attention in recent years. For $g = 2$, it is evident from Example 5.2.1 that $n \leq 7$ in such an array and Example 3.5.1 shows the attainability of this bound. For $g = 3$, Hedayat, Seiden, and Stufken (1997) showed that $n \leq 5$, which is sharper than the bound $n \leq 8$ given by Theorem 5.2.3. For $g \geq 4$, Hedayat, Stufken and Su (1997) showed that $n \leq g + 1$. As noted by these authors (see also Fujii et al., 1987), for both $g = 3$ and $g \geq 4$ the respective bounds are attainable, such as for $g \geq 4$, an $OA(2.3^g, g + 1, 3, g)$ can be constructed by repeating the rows of an $OA(3^g, g + 1, 3, g)$ given by Theorem 3.4.2.

5.3 LINEAR PROGRAMMING AND OTHER BOUNDS

In this section we primarily discuss some bounds that arose originally from coding theoretic considerations (see Delsarte, 1973; MacWilliams and Sloane, 1977). Let A be a symmetric orthogonal array $OA(N, n, m, g)$. The *Hamming distance* between any two rows of A is the number of places they differ; namely it equals n minus the number of coincidences between the two rows. For $1 \leq u \leq N, 0 \leq t \leq n$, let $a_t(u)$ be the number of rows of A having Hamming distance t with the uth row. Then the distance distribution of A is given by $\{a_0, a_1, \ldots, a_n\}$, where

$$a_t = \frac{1}{N} \sum_{u=1}^{N} a_t(u), \qquad 0 \leq t \leq n. \tag{5.3.1}$$

The distance enumerator for A is given by the polynomial

$$Q(z_1, z_2) = \sum_{t=0}^{n} a_t {z_1}^{n-t} {z_2}^{t}, \tag{5.3.2}$$

and the MacWilliams transform of the distance distribution is defined as $\{a_0{}^*, a_1{}^*, \ldots, a_n{}^*\}$, where

$$\frac{1}{N} Q(z_1 + (m-1)z_2, z_1 - z_2) = \sum_{s=0}^{n} a_s{}^* {z_1}^{n-s} {z_2}^{s}. \tag{5.3.3}$$

From (5.3.2) and (5.3.3), it is not hard to see that

$$a_s^* = \frac{1}{N} \sum_{t=0}^{n} K_s^{(m)}(t; n) a_t, \tag{5.3.4}$$

where $K_s^{(m)}(\cdot; n)$ is the Krawtchouk polynomial defined as

$$K_s^{(m)}(t; n) = \sum_{\nu=0}^{s} (-1)^{\nu} (m-1)^{s-\nu} \binom{t}{\nu} \binom{n-t}{s-\nu}.$$

Also, from (5.3.1), note that

$$N = \sum_{t=0}^{n} a_t \quad \text{and} \quad a_t \geq 0, \quad 0 \leq t \leq n. \tag{5.3.5}$$

The following useful lemma is originally due to Delsarte (1973). The notation developed in Chapter 2 however, helps us in sketching what we consider a more elementary proof.

Lemma 5.3.1. *(a)* $a_0 \geq 1$, *(b)* $a_0^* = 1$ *and* $a_s^* \geq 0$, $1 \leq s \leq n$, *and (c)* $a_1^* = \cdots = a_g^* = 0$.

Proof. Let Ω be the set of all binary n-tuples. For $\boldsymbol{x} \in \Omega$, define the matrices $P^{\boldsymbol{x}}$ and $W^{\boldsymbol{x}}$ as in (2.2.4) and (2.5.3), respectively, with $m_1 = \cdots = m_n = m$. Also, for $\boldsymbol{x} = x_1 \ldots x_n \in \Omega$, let

$$Z^{\boldsymbol{x}} = \bigotimes_{i=1}^{n} Z^{x_i}, \tag{5.3.6}$$

where

$$Z^{x_i} = \begin{cases} I & \text{if } x_i = 0, \\ J - I & \text{if } x_i = 1, 1 \leq i \leq n, \end{cases} \tag{5.3.7}$$

with I representing an identity matrix of order m and J, an $m \times m$ matrix of all ones. For $0 \leq t \leq n$, let

$$Z_{(t)} = \sum_{\boldsymbol{x} \in \Omega_{(t)}} Z^{\boldsymbol{x}}, \qquad W_{(t)} = \sum_{\boldsymbol{x} \in \Omega_{(t)}} W^{\boldsymbol{x}}, \tag{5.3.8}$$

where $\Omega_{(t)} (\subset \Omega)$ consists of binary n-tuples with exactly t components equal to unity.

For any scalars z_1, z_2, let $V = V(z_1, z_2)$ be the $m \times m$ matrix with each diagonal element equal to $z_1 + (m-1)z_2$ and each off-diagonal element equal to $z_1 - z_2$. Then

$$V = \{z_1 + (m-1)z_2\}I + (z_1 - z_2)(J - I) = z_1 J + z_2(mI - J).$$

Hence, writing the n-fold Kronecker product of V in two ways, it follows from (2.5.3), (2.5.4), and (5.3.6)–(5.3.8) that

$$\sum_{t=0}^{n} \{z_1 + (m-1)z_2\}^{n-t}(z_1 - z_2)^t Z_{(t)} = m^n \sum_{t=0}^{n} {z_1}^{n-t}{z_2}^t W_{(t)}. \qquad (5.3.9)$$

For $0 \le j_1, \ldots, j_n \le m-1$, let $r(j_1 \ldots j_n)$ be the number of times the n-tuple $j_1 \ldots j_n$ appears as a row of the orthogonal array under consideration. Define $\boldsymbol{r}$ as the $m^n \times 1$ vector with elements $r(j_1 \ldots j_n)$ arranged in lexicographic order. Then, from (5.3.1) and (5.3.6)–(5.3.8), it can be seen that

$$a_t = \frac{1}{N}\boldsymbol{r}' Z_{(t)} \boldsymbol{r}, \qquad 0 \le t \le n. \qquad (5.3.10)$$

Hence by (5.3.2) and (5.3.9),

$$Q(z_1 + (m-1)z_2, z_1 - z_2) = \frac{m^n}{N} \sum_{t=0}^{n} {z_1}^{n-t}{z_2}^t \boldsymbol{r}' W_{(t)} \boldsymbol{r},$$

and a comparison with (5.3.3) yields

$$a_s{}^* = \frac{m^n}{N^2}\boldsymbol{r}' W_{(s)} \boldsymbol{r}, \qquad 0 \le s \le n. \qquad (5.3.11)$$

Now, by (5.3.6)–(5.3.8) and (5.3.10), $a_0 = \boldsymbol{r}'\boldsymbol{r}/N$. Since the elements of $\boldsymbol{r}$ are integers, $\boldsymbol{r}'\boldsymbol{r} \ge \sum r(j_1 \ldots j_n) = N$, and hence (a) follows. From (5.3.4) and (5.3.5), noting that ${K_0}^{(m)}(t; n) \equiv 1$, we get $a_0{}^* = 1$. The truth of the rest of (b) follows from (5.3.8) and (5.3.11), noting that $W^{\boldsymbol{x}}$ is nonnegative definite for each $\boldsymbol{x}$. Finally, in order to prove (c), we note that if $1 \le s \le g$ then by (2.2.4) and (2.2.5), $P^{\boldsymbol{x}}\boldsymbol{r} = \boldsymbol{0}$ for every $\boldsymbol{x} \in \Omega_{(s)}$, since an orthogonal array of strength g is being considered (see the proof of Lemma 2.6.1). But, as noted in (2.5.2), $(P^{\boldsymbol{x}})'P^{\boldsymbol{x}} = W^{\boldsymbol{x}}$. Hence (5.3.8) yields $\boldsymbol{r}' W_{(s)} \boldsymbol{r} = 0$, $1 \le s \le g$ and from (5.3.11), part (c) follows. ■

As a consequence of (5.3.5) and Lemma 5.3.1, we get the following result:

Theorem 5.3.1. *Given positive integers n, m, g $(m \ge 2, n \ge g)$, consider the linear programming problem of choosing real numbers $a_0, a_1, \ldots, a_n$ so* *as to minimize*

$$a_0 + a_1 + \cdots + a_n \tag{5.3.12}$$

subject to

$$a_0 \geq 1, a_t \geq 0 (1 \leq t \leq n),$$
$$a_0{}^* = 1, a_s{}^* \geq 0 (1 \leq s \leq n), a_s{}^* = 0 (1 \leq s \leq g),$$

where $a_0{}^*, a_1{}^*, \ldots, a_n{}^*$ *are given by (5.3.4). Let* N_{LP} *be the minimal value of (5.3.12). Then any* $OA(N, n, m, g)$ *satisfies* $N \geq N_{LP}$. ■

Clearly, in an $OA(N, n, m, g)$, N must be an integral multiple of m^g. Hence, if N_{LP} is not divisible by m^g, then we can increase the lower bound to the next smallest multiple of m^g. Following Delsarte (1973), it can be seen that the linear programming bound is always at least as sharp as Rao's bound. Sloane and Stufken (1996) generalized the linear programming bound to the case of asymmetric orthogonal arrays and reported the following result:

Theorem 5.3.2. *Given positive integers* n_1, n_2, m_1, m_2, $g(m_1, m_2 \geq 2;$ $m_1 \neq m_2;$ $n_1 + n_2 \geq g)$, *choose real numbers* $a_{t_1 t_2} (0 \leq t_1 \leq n_1, 0 \leq t_2 \leq n_2)$ *so as to minimize*

$$\sum_{t_1=0}^{n_1} \sum_{t_2=0}^{n_2} a_{t_1 t_2} \tag{5.3.13}$$

subject to

$$a_{00} \geq 1, a_{t_1 t_2} \geq 0 (0 \leq t_1 \leq n_1, 0 \leq t_2 \leq n_2),$$
$$a_{00}{}^* = 1, a^*{}_{s_1 s_2} \geq 0 (0 \leq s_1 \leq n_1, 0 \leq s_2 \leq n_2, (s_1, s_2) \neq (0, 0)),$$
$$a^*{}_{s_1 s_2} = 0 (\text{ for } 1 \leq s_1 + s_2 \leq g),$$

where

$$a^*{}_{s_1 s_2} = \frac{1}{N} \sum_{t_1=0}^{n_1} \sum_{t_2=0}^{n_2} K_{s_1}^{(m_1)}(t_1; n_1) K_{s_2}^{(m_2)}(t_2; n_2) a_{t_1 t_2}.$$

Let N_{LP} *be the minimal value of (5.3.13). Then any* $OA(N, n_1 + n_2, m_1{}^{n_1} \times m_2{}^{n_2}, g)$ *satisfies* $N \geq N_{LP}$. ■

The number of rows in an $OA(N, n_1 + n_2, m_1^{n_1} \times m_2^{n_2}, g)$ is divisible by

$$N_0 = \text{l.c.m.}\{m_1^{i_1} m_2^{i_2} : 0 \le i_1 \le n_1, \quad 0 \le i_2 \le n_2, 0 \le i_1 + i_2 \le g\}, \tag{5.3.14}$$

where l.c.m. stands for the lowest common multiple. Hence if N_{LP} is not divisible by N_0 then the lower bound in Theorem 5.3.2 can be increased to the next smallest multiple of N_0. Sloane and Stufken (1996) numerically evaluated the linear programming bound for (i) $m_1 = 2$, $m_2 = 3$, $g = 2, 3, 4$ and (ii) $m_1 = 2$, $m_2 = 4$, $g = 2$, over practically useful ranges of n_1 and n_2, and obtained improvement over Rao's bound in many situations. We present below an illustrative example and refer to the original paper for further details.

Example 5.3.1. Consider an $OA(N, n_1 + n_2, 2^{n_1} \times 3^{n_2}, 3)$. Define N_0 as in (5.3.14). Let N^*_{LP} equal N_{LP} if N_{LP} is an integral multiple of N_0; otherwise, let N^*_{LP} be obtained by rounding N_{LP} up to the next multiple of N_0. Similarly let N^*_{Rao} be defined with reference to Rao's bound. Sloane and Stufken (1996) noted that over the range $n_1 + 2n_2 \le 75$, $n_1 \le 60$, N^*_{LP} and N^*_{Rao} agree, *except* in the following cases:

(i) $n_1 = 1, \quad n_2 = 9u, \quad u \ge 1$ when $N^*_{LP} = N^*_{\text{Rao}} + 54$.
(ii) $n_2 = 2, \quad n_1 = 24u + 21, \quad u \ge 0$, when $N^*_{LP} = N^*_{\text{Rao}} + 72$.
(iii) $n_1 = 2u + 1, \quad n_2 = 36 - u$, for $1 \le u \le 29, \quad u \ne 12, 18$, when $N^*_{LP} = 432$ but $N^*_{\text{Rao}} = 216$. ■

With heavier notation, Theorem 5.3.2 can be extended in an obvious manner to asymmetric orthogonal arrays of the type $OA(N, \sum n_i, m_1^{n_1} \times \cdots \times m_u^{n_u}, g)$, where $m_1, \ldots, m_u$ are distinct and $u \ge 3$. For further results on coding theory, with potential applicability in proving nonexistence of orthogonal arrays, see Fujii (1976), MacWilliams and Sloane (1977), and the references in Sloane and Stufken (1996).

Before concluding this section, we indicate certain results obtained by Wang and Wu (1992) via a complete enumeration of all possibilities. These authors found that

(a) in an $OA(12, n + 1, 3 \times 2^n, 2), n \le 4$;
(b) in an $OA(24, n + 1, 3 \times 2^n, 2), n \le 16$;
(c) in an $OA(20, n + 1, 5 \times 2^n, 2), n \le 8$.

By actual construction, they also showed that these bounds are attainable.

It is an interesting but difficult problem to obtain a general result which covers (a)–(c) above as special cases. To that effect, we have the following conjecture (see Dey and Midha, 1996):

If $m(\geq 3)$ is odd and $t \geq 1$ then in an $OA(4mt, n+1, m \times 2^n, 2)$, $n \leq 4mt - 2m - 2$.

5.4 ON THE TIGHT AND NEARLY TIGHT CASES

Let A be an orthogonal array $OA(N, n, m_1 \times \cdots \times m_n, g)$ where $g(= 2f, f \geq 1)$ is *even*. As in the last section, let $r(j_1 \ldots j_n)$ be the frequency with which the n-tuple $j_1 \ldots j_n$ appears as a row of $A (0 \leq j_i \leq m_i - 1, 1 \leq i \leq n)$. Let $v = \prod m_i$ and define R as the diagonal matrix of order v with diagonal elements $r(j_1 \ldots j_n)$ arranged in the lexicographic order. Then following the proof of Theorem 2.6.1 (see (2.6.11)),

$$P^{(1)} R (P^{(1)})' = \left(\frac{N}{v}\right) I_{\alpha_f}, \tag{5.4.1}$$

where α_f is given by (2.3.11) and the $\alpha_f \times v$ matrix $P^{(1)}$ is as defined in (2.3.9).

In view of (2.2.4) and (2.3.9), the columns of $P^{(1)}$ can be labelled by n-tuples $j_1 \ldots j_n$ in the lexicographic order $(0 \leq j_i \leq m_i - 1, 1 \leq i \leq n)$. Let $\pi(j_1 \ldots j_n)$ be the $(j_1 \ldots j_n)$th column of $P^{(1)}$ and G be an $\alpha_f \times N$ matrix such that, for $1 \leq i \leq N$, if the ith row of the array A is given by the n-tuple $u_1 \ldots u_n$ then the ith column of G equals $\pi(u_1 \ldots u_n)$. Then clearly, $GG' = P^{(1)} R (P^{(1)})'$ so that by (5.4.1),

$$GG' = \left(\frac{N}{v}\right) I_{\alpha_f}. \tag{5.4.2}$$

Now suppose that the array A is tight in the sense of attaining Rao's bound, namely $N = \alpha_f$. Then G is a square matrix and hence by (5.4.2),

$$G'G = \left(\frac{N}{v}\right) I_{\alpha_f}. \tag{5.4.3}$$

The simple identity (5.4.3), implicit in the work of Delsarte (1973), is of great help in studying the existence of tight orthogonal arrays and forms the foundation of the Delsarte theory for such arrays.

We now consider in some detail the case where the array A has strength *two*. Then $f = 1$ and by (2.2.4), (2.2.5) and (2.3.9),

$$P^{(1)} = \begin{bmatrix} v^{-1/2}\mathbf{1}'_{m_1} \otimes \mathbf{1}'_{m_2} \otimes \cdots \otimes \mathbf{1}'_{m_n} \\ \left(\dfrac{v}{m_1}\right)^{-1/2} P_1 \otimes \mathbf{1}'_{m_2} \otimes \cdots \otimes \mathbf{1}'_{m_n} \\ \vdots \\ \left(\dfrac{v}{m_n}\right)^{-1/2} \mathbf{1}'_{m_1} \otimes \mathbf{1}'_{m_2} \otimes \cdots \otimes P_n \end{bmatrix}, \tag{5.4.4}$$

where $P_1, \ldots, P_n$ are as defined in (2.2.2), and as before, $\mathbf{1}_\nu$ is the $\nu \times 1$ vector of all ones. By (2.2.2) and (5.4.4), for any $j_1, \ldots, j_n, u_1, \ldots, u_n (0 \leq j_i, u_i \leq m_i - 1, 1 \leq i \leq n)$,

$$\pi(j_1 \ldots j_n)'\pi(u_1 \ldots u_n) = \frac{1 + \sum_{i=1}^{n}\{m_i\delta(j_i, u_i) - 1\}}{v}, \tag{5.4.5}$$

where $\delta(j_i, u_i)$ is the Kronecker's delta, which equals 1 if $j_i = u_i$ and zero otherwise. From (5.4.3) and (5.4.5) it follows that if two distinct rows of the array A are given by n-tuples $j_1 \ldots j_n$ and $u_1 \ldots u_n$, then

$$1 + \sum_{i=1}^{n}\{m_i\delta(j_i, u_i) - 1\} = 0. \tag{5.4.6}$$

Mukerjee and Wu (1995) examined the use of (5.4.6) in studying the nonexistence of tight asymmetric orthogonal arrays of strength two. In order to illustrate their approach, we consider a tight $OA(N, n_1 + n_2, {m_1}^{n_1} \times {m_2}^{n_2}, 2)$ where $m_1 \neq m_2$. With reference to any two distinct rows of the array, let $h^{(1)}$ and $h^{(2)}$ be the number of coincidences arising from the m_1-symbol and m_2-symbol columns, respectively. Then

$$0 \leq h^{(i)} \leq n_i, \qquad i = 1, 2, \tag{5.4.7}$$

and by (5.4.6),

$$m_1 h^{(1)} + m_2 h^{(2)} = n_1 + n_2 - 1. \tag{5.4.8}$$

One can enumerate all possible integral valued solutions to (5.4.8) over the range (5.4.7) and thus identify the set H of possible values of $h^{(1)}$. Consider now the first row of the subarray given by the m_1-symbol columns. Among the other rows of the subarray, let there be p_h rows having h coincidences with the first row, $h \in H$. Then

$$\sum_{h \in H} p_h = N - 1, \tag{5.4.9}$$

and following (5.2.3),

$$\sum_{h \in H} h p_h = n_1 \left(\frac{N}{m_1} - 1 \right), \quad \sum_{h \in H} \binom{h}{2} p_h = \binom{n_1}{2} \left(\frac{N}{{m_1}^2} - 1 \right). \tag{5.4.10}$$

If the system of equations given by (5.4.9) and (5.4.10) does not admit a nonnegative integral valued solution for $p_h, h \in H$, then the nonexistence of the tight $OA(N, n_1 + n_2, {m_1}^{n_1} \times {m_2}^{n_2}, 2)$ follows. In fact, as illustrated in Example 5.4.1 below, even when this system of equations admits nonnegative integral valued solution(s), one may be able to prove nonexistence by a careful examination of the nature of the solution(s). As noted in Mukerjee and Wu (1995), if one works with $h^{(2)}$ instead of $h^{(1)}$, then the resulting system of equations becomes equivalent to (5.4.9) and (5.4.10) and hence yields identical results.

Example 5.4.1. Consider a tight $OA(100, n_1 + n_2, 2^{n_1} \times 5^{n_2}, 2)$, where n_1 and n_2 are positive integers satisfying (see Theorem 2.6.2)

$$n_1 + 4n_2 = 99. \tag{5.4.11}$$

Application of Theorem 5.2.2 to the 5-symbol subarray yields $n_2 \leq 23$. Consider the case $n_2 = 23$. Then by (5.4.11), $n_1 = 7$, and from (5.4.7), (5.4.8), the only possible values of $h^{(1)}$ are 2 and 7. The equations (5.4.9) and (5.4.10) now reduce to

$$p_2 + p_7 = 99, \quad 2p_2 + 7p_7 = 343, \quad p_2 + 21p_7 = 504,$$

which are inconsistent. Hence nonexistence follows for $n_2 = 23$. Next let $n_2 = 22$. Then by (5.4.11), $n_1 = 11$, and as before, the only possible values of $h^{(1)}$ are 1, 6 and 11. The equations (5.4.9) and (5.4.10) become

$$p_1 + p_6 + p_{11} = 99, \quad p_1 + 6p_6 + 11p_{11} = 539, \quad 15p_6 + 55p_{11} = 1320,$$

with unique solution $p_1 = 11$, $p_6 = 88$, $p_{11} = 0$. Without loss of generality, suppose that the first row of the two-symbol subarray is $00\ldots0$. As $p_1 = 11$, there are 11 rows of this subarray having exactly one zero and ten unities. By rearranging columns, if necessary, let the first of these 11 rows be, say $e_1 = 011\ldots1$. The second, say e_2, of these 11 rows must also have one zero and 10 unities. This, however, is impossible, since, as noted above, e_1 and e_2 must have either 1 or 6 coincidences. Therefore nonexistence follows for $n_2 = 22$ as well.

For $n_2 = 21$ one can prove nonexistence in a similar manner by checking that the equations (5.4.9) and (5.4.10) do not admit a nonnegative integral valued solution. Thus in a tight $OA(100, n_1 + n_2, 2^{n_1} \times 5^{n_2}, 2)$, we have $n_2 \leq 20$. ■

Remark 5.4.1. More generally, Mukerjee and Wu (1995) used (5.4.7)–(5.4.10) to show that $n_2 \leq 4m$ in a tight $OA(4m^2, n_1 + n_2, 2^{n_1} \times m^{n_2}, 2)$, where $m(\geq 3)$ is odd and, as in (5.4.11), n_1 and n_2 satisfy $n_1 + (m-1)n_2 = 4m^2 - 1$. Does such an array exist for $n_2 = 4m$ and hence $n_1 = 4m - 1$? The answer to this question is in the affirmative if $m(\geq 3)$ is an odd prime or prime power and a Hadamard matrix of order $4m$ exists, for then, by Theorem 3.2.2 and Remark 3.5.3, a symmetric $OA(4m, 4m-1, 2, 2)$, say B, and a difference matrix $D(4m, 4m; m)$ are available. So one can proceed as in Section 4.4 to get a tight $OA(4m^2, 8m-1, 2^{4m-1} \times m^{4m}, 2)$ as

$$[\mathbf{0}_m \odot B \vdots \boldsymbol{\lambda}_m \odot D(4m, 4m; m)],$$

where $\odot$ stands for Kronecker sum, $\mathbf{0}_m$ is the $m \times 1$ vector of zeros and $\boldsymbol{\lambda}_m = (\rho_0, \rho_1, \ldots, \rho_{m-1})'$, with $\rho_0, \ldots, \rho_{m-1}$ representing the distinct elements of $GF(m)$. ■

Remark 5.4.2. From (5.4.7)–(5.4.10), Mukerjee and Wu (1995) also deduced that in the setup of Theorem 4.6.1, if $p = 1$, then

$$n_2 \leq \frac{m^r - m^{k+1}}{m^k - 1} + 1.$$

This shows that for $p = 1$ the construction of Wu et al. (1992), as summarized in Theorem 4.6.1, cannot be improved upon in so far as the accommodation of more m^k-level columns is concerned. ■

It is sometimes possible to conclude the nonexistence of nearly tight orthogonal arrays from that of tight arrays. The following result, due to Mukerjee and Wu (1995), is helpful for this purpose:

Theorem 5.4.1.

(a) *The existence of an $OA(N, n, m_1 \times \cdots \times m_n, 2)$, which is nearly tight in the sense that*

$$\sum_{i=1}^{n}(m_i - 1) = N - 2, \tag{5.4.12}$$

implies the existence of a tight $OA(N, n+1, 2 \times m_1 \times \cdots \times m_n, 2)$.

(b) *Suppose that* N *is not an integral multiple of 3. Then the existence of an* $OA(N, n, m_1 \times \cdots \times m_n, 2)$, *which is nearly tight in the sense that*

$$\sum_{i=1}^{n}(m_i - 1) = N - 3, \tag{5.4.13}$$

implies the existence of a tight $OA(N, n+2, 2^2 \times m_1 \times \cdots \times m_n, 2)$.

Proof. Given an $OA(N, n, m_1 \times \cdots \times m_n, 2)$, say A, define the matrix G as in the context of (5.4.2). Then G has $1 + \sum_{i=1}^{n}(m_i - 1)$ rows and N columns and satisfies (5.4.2). In view of (5.4.4), each element in the first row of G equals $v^{-1/2}$.

(a) Suppose that A is nearly saturated in the sense of (5.4.12). Then G is $(N-1) \times N$, and hence, by (5.4.2), there exists an $N \times 1$ vector $\boldsymbol{q} = (q_1, \ldots, q_N)'$ such that the $N \times N$ matrix

$$\begin{bmatrix} \boldsymbol{q}' \\ \left(\frac{v}{N}\right)^{1/2} G \end{bmatrix}$$

is orthogonal. Then

$$\boldsymbol{q}\boldsymbol{q}' + \left(\frac{v}{N}\right) G'G = I_N.$$

Equating diagonal elements from both sides of the above, by (5.4.5) and (5.4.12),

$$q_i^2 + \frac{N-1}{N} = 1 \Rightarrow \qquad q_i = \pm N^{-1/2}, \qquad 1 \le i \le N. \tag{5.4.14}$$

As $\boldsymbol{q}'$ is orthogonal to the first row of G, it follows from (5.4.14) that half the elements of $\boldsymbol{q}$ equal $N^{-1/2}$, while the remaining half equal $-N^{-1/2}$ (i.e., N must be even). Let $\bar{A}$ be an array obtained by adding a two-symbol column to A such that for $1 \le i \le N$, the ith entry of this column equals zero if $q_i = -N^{-1/2}$ and 1 if $q_i = N^{-1/2}$. We claim that $\bar{A}$ is an $OA(N, n+1, 2 \times m_1 \times \cdots \times m_n, 2)$.

To that effect, it must be shown that the newly added two-symbol column is orthogonal to each of the columns of A. Without loss of generality, consider the first column of A, say $(l_1, \ldots, l_N)'$, where $0 \leq l_i \leq m_1 - 1$ for each i. For $u = 0, 1$ and $0 \leq j \leq m_1 - 1$, let ϕ_{uj} be the frequency of occurrence of the pair (u, j) as a row in the $N \times 2$ subarray of $\bar{A}$ given by the newly added two-symbol column and the first column of A. Since $\boldsymbol{q}'$ is orthogonal to each row of G, by (5.4.4) and the definition of G,

$$\sum_{i=1}^{N} q_i \boldsymbol{\pi}_1(l_i) = \mathbf{0}, \tag{5.4.15}$$

where $\boldsymbol{\pi}_1(j)$ is the jth column of P_1 $(0 \leq j \leq m_1 - 1)$ and $\mathbf{0}$ is the null vector of order $m_1 - 1$. By our construction, for $0 \leq j \leq m_1 - 1$, the pair (q_i, l_i) equals $(-N^{-1/2}, j)$ for ϕ_{0j} choices of i and $(N^{-1/2}, j)$ for ϕ_{1j} choices of i. Hence by (5.4.15),

$$\sum_{j=0}^{m_1-1} (\phi_{1j} - \phi_{0j}) \boldsymbol{\pi}_1(j) = \mathbf{0}.$$

Therefore, by (2.2.2), there exists a constant ϕ such that $\phi_{1j} - \phi_{0j} = \phi$ for each j. Recalling the structure of $\boldsymbol{q}$, $\sum_{j=0}^{m_1-1} \phi_{0j} = \sum_{j=0}^{m_1-1} \phi_{1j} = N/2$. Hence $\phi = 0$; that is, $\phi_{0j} = \phi_{1j}$ for each j. At the same time, since A is an orthogonal array of strength two, $\phi_{0j} + \phi_{1j} = N/m_1$ for each j. Thus $\phi_{0j} = \phi_{1j} = N/(2m_1)$ for each j, and our claim is established.

(b) Suppose that A is nearly saturated in the sense of (5.4.13). Then G is $(N-2) \times N$ and hence, by (5.4.2), there exists an $N \times 2$ matrix $Q = (\boldsymbol{q}_1, \ldots, \boldsymbol{q}_N)'$ such that

$$\boldsymbol{q}_1' = (w, 0), \tag{5.4.16}$$

for some $w \geq 0$, and the $N \times N$ matrix

$$\begin{bmatrix} Q' \\ \left(\frac{v}{N}\right)^{1/2} G \end{bmatrix}$$

is orthogonal. Then

$$QQ' + \left(\frac{v}{N}\right) G'G = I_N.$$

Equating corresponding elements from both sides of the above, by (5.4.5) and (5.4.13),

$$\boldsymbol{q}_i'\boldsymbol{q}_i + \frac{N-2}{N} = 1 \Rightarrow \qquad \boldsymbol{q}_i'\boldsymbol{q}_i = \frac{2}{N}, 1 \le i \le N, \quad (5.4.17)$$

$$\boldsymbol{q}_i'\boldsymbol{q}_j + \frac{\mu_{ij}}{N} = 0, \qquad 1 \le i \ne j \le N, \quad (5.4.18)$$

where μ_{ij}'s are integers. By (5.4.17), (5.4.18) and the Cauchy-Schwarz inequality,

$$|\mu_{ij}| \le 2, \qquad 1 \le i \ne j \le N. \quad (5.4.19)$$

Since $\boldsymbol{q}_1'$ is as in (5.4.16) with $w \ge 0$, by (5.4.17), $\boldsymbol{q}_1' = c(1, 0)$, where $c = \sqrt{2/N}$. Hence by (5.4.18) and (5.4.19), the only possibilities for $\boldsymbol{q}_i'$, $2 \le i \le N$, are

$$c(0, \pm 1), \quad (5.4.20a)$$

$$c(\pm\frac{1}{2}, \pm\frac{\sqrt{3}}{2}), \quad (5.4.20b)$$

$$c(\pm 1, 0). \quad (5.4.20c)$$

However, by (5.4.18) noting that μ_{ij}'s are integers, Q cannot simultaneously have two rows one of which is of the form (5.4.20a) and the other of the form (5.4.20b). Hence two cases arise:

Case 1. Each row of Q is of the form $c(0, 1)$, or $c(0, -1)$, or $c(1, 0)$, or $c(-1, 0)$.

Case 2. Each row of Q is of the form $c(\frac{1}{2}, \frac{\sqrt{3}}{2})$, or $c(\frac{1}{2}, \frac{-\sqrt{3}}{2})$, or $c(-\frac{1}{2}, \frac{\sqrt{3}}{2})$, or $c(-\frac{1}{2}, \frac{-\sqrt{3}}{2})$, or $c(1, 0)$, or $c(-1, 0)$.

Considering Case 2 first, suppose that the vectors listed under this case appear as rows of Q with respective frequencies $\phi_1, \ldots, \phi_6$. By the definition of Q, its second column has length unity. Hence $\phi_1 + \phi_2 + \phi_3 + \phi_4 = 2N/3$, which is impossible as N is not an integral multiple of 3. Thus the rows of Q must be as in Case 1. Then let the vectors listed under Case 1 appear as rows of Q with respective frequencies $\phi_1{}^*, \ldots, \phi_4{}^*$. Since each column of Q has length unity and is orthogonal to the first column of G',

$$\phi_3{}^* + \phi_4{}^* = \phi_1{}^* + \phi_2{}^* = \frac{N}{2}, \quad \phi_3{}^* - \phi_4{}^* = \phi_1{}^* - \phi_2{}^* = 0,$$

so that $\phi_1{}^* = \phi_2{}^* = \phi_3{}^* = \phi_4{}^* = N/4$ (i.e., N is an integral multiple of 4).

We now add a two-symbol column to A such that for $1 \le i \le N$, the ith entry of this column equals 1 if $\boldsymbol{q}_i' = c(0, 1)$ or $c(1, 0)$ and zero if $\boldsymbol{q}_i' = c(0, -1)$ or $c(-1, 0)$. As in the proof of part (a), this yields an $OA(N, n+1, 2 \times m_1 \times \cdots \times m_n, 2)$ which, by (5.4.13), is nearly saturated in the sense of (5.4.12). Application of part (a) now guarantees the existence of an $OA(N, n+2, 2^2 \times m_1 \times \cdots \times m_n, 2)$. ■

Remark 5.4.3. The conclusion of Theorem 5.4.1(b) does not hold when N is an integral multiple of 3. To see this, consider an $OA(18, 8, 2 \times 3^7, 2)$, which can be constructed, following Section 4.4, as

$$[\mathbf{0}_3 \odot B \,\vdots\, \boldsymbol{\lambda}_3 \odot D(6, 6; 3)],$$

where $\mathbf{0}_3 = (0, 0, 0)'$,

$$B = \begin{bmatrix} 0 & 0 & 0 & 1 & 1 & 1 \\ 0 & 1 & 2 & 0 & 1 & 2 \end{bmatrix}',$$

$\boldsymbol{\lambda}_3 = (0, 1, 2)'$ and $D(6, 6; 3)$ is as shown in (3.5.1). This array is nearly tight in the sense of Theorem 5.4.1(b). However, as 18 is not an integral multiple of 4, it is not possible to add a two-symbol column to this array and retain orthogonality. ■

Example 5.4.2. It was seen earlier in Example 5.4.1 that a tight $OA(100, n_1 + n_2, 2^{n_1} \times 5^{n_2}, 2)$, where

$$n_1 + 4n_2 = 99, \quad 21 \le n_2 \le 23, \tag{5.4.21}$$

is nonexistent. Hence by Theorem 5.4.1, it follows that the nearly tight arrays $OA(100, n_1+n_2-1, 2^{n_1-1} \times 5^{n_2}, 2)$ and $OA(100, n_1+n_2-2, 2^{n_1-2} \times 5^{n_2}, 2)$ do not exist whenever n_1 and n_2 satisfy (5.4.21). ■

Consider now a tight $OA(N, n_1 + n_2, m_1{}^{n_1} \times m_2{}^{n_2}, 4)$. From (5.4.3), proceeding as with (5.4.8), one can check that such an array must satisfy

$$\begin{aligned} 1 - \tfrac{3}{2}n + \tfrac{1}{2}n^2 + \tfrac{1}{2}m_1{}^2 h^{(1)}(h^{(1)} - 1) + \tfrac{1}{2}m_2{}^2 h^{(2)}(h^{(2)} - 1) \\ + m_1 m_2 h^{(1)} h^{(2)} - (n-2)(m_1 h^{(1)} + m_2 h^{(2)}) = 0, \end{aligned} \tag{5.4.22}$$

where $n = n_1 + n_2$ and $h^{(1)}$ and $h^{(2)}$ are as in the context of (5.4.8). Mukerjee and Wu (1995) studied integral valued solutions of (5.4.22) for $h^{(1)}$ and $h^{(2)}$ subject to (5.4.7) and used the fact that N must be an integral multiple of N_0 as defined in (5.3.14) to conclude that a tight $OA(N, n_1+n_2, m_1{}^{n_1} \times m_2{}^{n_2}, 4)$ does not exist over the range

$$N \leq 1000, \quad n_1 \geq 1, n_2 \geq 1, \quad n_1 + n_2 \geq 5, \quad 2 \leq m_1 < m_2 \leq 7.$$

As for symmetric orthogonal arrays, equation (5.4.3) (or, equivalently, its coding theoretic counterpart as in Delsarte, 1973) has led to a very rich theory. Even for arrays of odd strength, results are available via Theorem 3.3.1 and Lemma 5.2.2. The proofs are deep and make extensive use of association algebras and number theoretic considerations. We only summarize below the main results.

Theorem 5.4.2.

(a) *(Noda, 1986). If there exists a tight $OA(N, n, m, 3)$, then either*

- ***(i)*** $(N, n, m) = (2n, n, 2)$ *with* $n \equiv 0 (\text{mod } 4)$, *or*
- ***(ii)*** $(N, n, m) = (m^3, m + 2, m)$ *with* $m (\geq 4)$ *even.*

(b) *(Noda, 1979). If there exists a tight $OA(N, n, m, 4)$, then either*

- ***(i)*** $(N, n, m) = (16, 5, 2)$, *or*
- ***(ii)*** $(N, n, m) = (243, 11, 3)$, *or*
- ***(iii)*** $(N, n, m) = (\frac{9}{2}u^2(9u^2 - 1), \frac{1}{5}(9u^2 + 1), 6)$, *for some integer* u *with* $u \equiv 0 (\text{mod } 3)$, $u \equiv \pm 1 (\text{mod } 5)$ *and* $u \equiv 5 (\text{mod } 16)$.

(c) *(Noda, 1986). If there exists a tight $OA(N, n, m, 5)$, then either*

- ***(i)*** $(N, n, m) = (32, 6, 2)$, *or*
- ***(ii)*** $(N, n, m) = (729, 12, 3)$. ■

Theorem 5.4.3. *(Hong, 1986). For $g \geq 6$ and $m \geq 3$ there does not exist a tight $OA(N, n, m, g)$.* ■

Theorem 5.4.4. *(Mukerjee and Kageyama, 1994).*

(a) *A tight $OA(N, n, 2, 6)$ exists if and only if $n = 7$ or 23.*

(b) *A tight $OA(N, n, 2, 7)$ exists if and only if $n = 8$ or 24.*

(c) *For $8 \leq g \leq 13$ and $n \leq 10^9$, a tight $OA(N, n, 2, g)$ exists if and only if $n = g + 1$.* ■

Remark 5.4.4. It is of interest to examine the construction of tight arrays when existence is not ruled out by Theorems 5.4.2 and 5.4.4. The arrays in Theorem 5.4.2 (b (i)) or (c(i)) are given by Theorem 3.4.2 with $m = 2$ and $g = 4$ or 5, respectively. Similarly, Theorem 3.4.2 yields the arrays corresponding to $n = 7, 8$, and $g + 1$ in parts (a), (b), and (c) respectively of Theorem 5.4.4. The array corresponding to $n = 23$ in Theorem 5.4.4 (a) is given by the binary Golay code (MacWilliams and Sloane, 1977), and then the array corresponding to $n = 24$ in Theorem 5.4.4 (b) can be obtained via Theorem 3.3.1. The arrays in Theorem 5.4.2 (b(ii)) and (c(ii)) are given by the ternary and extended ternary Golay codes (MacWilliams and Sloane, 1977). The array in Theorem 5.4.2 (a(i)) can be obtained from Corollary 3.3.1 provided that a Hadamard matrix of order n exists, while that in Theorem 5.4.2 (a(ii)) can be constructed via Theorem 3.4.1 provided m is a power of 2. Finally it seems that the existence question of an array as in Theorem 5.4.2 (b(iii)) is still open. ■

EXERCISES

5.1. Verify (5.2.3).

5.2. Verify that the conclusion of Example 5.2.1 follows from Theorem 5.2.2 as well.

5.3. Verify (5.3.4) and (5.3.10).

5.4. Consider an $OA(4t, n+1, (2t) \times 2^n, 2)$, where $t (\geq 1)$ is an odd integer. Show that $n \leq 2$ and that this upper bound on n is attainable. [*Hint*: See Wang and Wu (1992).]

5.5. Verify the statement made in the first sentence of Remark 5.4.1.

5.6. Construct another example in the spirit of Remark 5.4.3.

5.7. Verify (5.4.22).

5.8. Use Theorem 3.3.1 and Lemma 5.2.2 to show that part (b) of Theorem 5.4.4 is implied by its part (a).

CHAPTER 6

More on Optimal Fractional Plans and Related Topics

6.1 INTRODUCTION AND PRELIMINARIES

We now return to the setup of Chapter 2. With reference to an $m_1 \times \cdots \times m_n$ factorial, assume the absence of factorial effects involving $f+1$ or more factors, and suppose that interest lies in the general mean and effects involving f or less factors ($1 \leq f \leq n-1$). This amounts to considering the model (2.3.12) with $f = t$, and the parametric vector of interest is $\boldsymbol{\beta}^{(1)}$ as defined in (2.3.8). Recall that $\boldsymbol{\beta}^{(1)}$ represents the general mean and complete sets of orthonormal contrasts belonging to factorial effects involving f or less factors. As in Section 2.5, let $\mathcal{D}_N$ be the class of N-run plans and $\mathcal{D}_N{}^*$ be a subclass of $\mathcal{D}_N$ consisting of those plans that keep $\boldsymbol{\beta}^{(1)}$ estimable when effects involving $f+1$ or more factors are absent. With α_f defined as in (2.3.11), by Theorem 2.3.2, $\mathcal{D}_N{}^*$ is nonempty if and only if $N \geq \alpha_f$, a condition that is assumed to hold throughout this chapter. Following the explanation provided in Section 2.4, we continue to refer to any member of $\mathcal{D}_N{}^*$ as a Resolution (f, f) plan, although, as noted there, such a member has often been called a Resolution $(2f+1)$ plan in the literature. In particular, a Resolution $(1, 1)$ plan is also known as a main effect plan.

Let $v = \prod m_i$ and, for any $d \in \mathcal{D}_N$, R_d be a $v \times v$ diagonal matrix with diagonal elements $r_d(j_1 \ldots j_n)$ arranged lexicographically, where $r_d(j_1 \ldots j_n)$ is the frequency of occurrence of the treatment combination $j_1 \ldots j_n$ in d. Also let the $\alpha_f \times v$ matrix $P^{(1)}$ be defined as in (2.3.9). The columns of $P^{(1)}$ correspond to the lexicographically ordered treatment combinations, a fact that will often be needed in this chapter. By (2.3.17), the information matrix for $\boldsymbol{\beta}^{(1)}$ under the plan d is

$$\mathcal{I}_d = P^{(1)} R_d (P^{(1)})' \qquad (6.1.1)$$

which is of order $\alpha_f \times \alpha_f$. For any $d \in \mathcal{D}_N$, by (6.1.1), proceeding exactly as in the proof of Lemma 2.5.1,

$$\operatorname{tr}(\mathcal{I}_d) = \frac{N\alpha_f}{v}. \tag{6.1.2}$$

By Theorem 2.6.1, if an orthogonal array $OA(N, n, m_1 \times \cdots \times m_n, 2f)$ exists, then it represents a universally optimal Resolution (f, f) plan in $\mathcal{D}_N$. The issue of optimality can become significantly more challenging when the nature of $N, n, m_1, \ldots, m_n$ and f rules out the availability of such an array. The corresponding developments are reviewed in this chapter.

Theorem 2.6.1 heuristically suggests that the plans given by structures that are close to orthogonal arrays in some sense should behave well from the point of view of optimality. Considerable work has been done in this direction, and Sections 6.2–6.4 summarize these results. Sections 6.2 and 6.3 explore the consequence of adding one or more run(s) to an orthogonal array, while Section 6.4 deals with arrays that are "nearly" orthogonal in the sense of Wang and Wu (1992). In Section 6.5 we discuss the connection between optimal weighing designs and optimal fractions of 2^n factorials. Optimal saturated or nearly saturated plans with two or three factors are reviewed in Section 6.6. Finally, Section 6.7 briefly reviews certain other plans that can be of interest, though they are not necessarily optimal.

6.2 AUGMENTED ORTHOGONAL ARRAYS: ADDITION OF ONE RUN

In this section we examine the optimality of plans obtained by adding a single run to an orthogonal array. Work in this area can be traced back to Mood (1946) and Mitchell (1974) though it was Cheng (1980a) who first systematically studied the problem for 2^n factorials. Cheng (1980a) worked under a very general class of optimality criteria as described below.

For any plan $d \in \mathcal{D}_N$, denote the eigenvalues of $\mathcal{I}_d$ by $\mu_{d1}, \ldots, \mu_{d\alpha_f}$. Then, in view of (6.1.2), we have

$$\mu_{di} \in \left[0, \frac{N\alpha_f}{v}\right], \qquad 1 \le i \le \alpha_f.$$

Following Cheng (1980a), an optimality criterion of type 1 is given by

$$\phi(\mathcal{I}_d) = \sum_{i=1}^{\alpha_f} q(\mu_{di}), \tag{6.2.1}$$

where q is a real-valued function defined on $[0, N\alpha_f/v]$ such that

(a) q is continuous, strictly convex and strictly decreasing on $[0, N\alpha_f/v]$; we include here the possibility that $\lim_{\mu \to 0^+} q(\mu) = q(0) = +\infty$;

(b) q is continuously differentiable and the derivative function q' is strictly concave on $(0, N\alpha_f/v)$.

Any generalized criterion of type 1 is the pointwise limit of a sequence of type 1 criteria. The class of generalized criteria of type 1 is very rich. It is easily seen that the D- and A-criteria, introduced in Section 2.5, are of type 1 (and hence of generalized type 1) with $q(\mu) = -log\mu$ and $q(\mu) = \mu^{-1}$, respectively. Also the E-criterion, introduced in the same section, is of generalized type 1.

A sufficient condition for a plan to be optimal with respect to (the minimization of) any generalized type 1 criterion was given by Cheng (1980a). Since, by (6.1.2), $\text{tr}(\mathcal{I}_d)$ is constant over $\mathcal{D}_N$, the condition may be stated as follows in the present context:

Lemma 6.2.1. *Let there exist a plan $d_0 \in \mathcal{D}_N$ such that*

(i) *$\mathcal{I}_{d_0}$ has two distinct eigenvalues, the larger of which has multiplicity unity,*

(ii) *$tr(\mathcal{I}_{d_0}{}^2) < \{tr(\mathcal{I}_{d_0})\}^2/(\alpha_f - 1)$, and*

(iii) *d_0 minimizes $tr(\mathcal{I}_d{}^2)$ over $\mathcal{D}_N$.*

Then d_0 is optimal in $\mathcal{D}_N$ with respect to every generalized type 1 criterion. ■

Remark 6.2.1. The proof of Lemma 6.2.1 is highly nontrivial and can be found in Cheng (1978). Observe that if d_0 satisfies the first two conditions of Lemma 6.2.1, then $\mathcal{I}_{d_0}$ must be positive definite, for otherwise, by (6.1.2) and condition (i), its eigenvalues are zero and $N\alpha_f/v$ with respective multiplicities $\alpha_f - 1$ and 1. So condition (ii) is violated. Thus any plan d_0, satisfying (i) and (ii) of Lemma 6.2.1, must be a Resolution(f, f) plan (see Theorem 2.3.1). ■

Remark 6.2.2. Cheng (1980a) also considered generalized criteria of type 2. These are, however, not relevant in the present context and hence will not be considered in the sequel. ■

Lemma 6.2.2. *Let $d_0(\in \mathcal{D}_N)$ be obtained by adding any treament combination to a plan given by an orthogonal array $OA(N-1, n, m_1 \times \cdots \times m_n, 2f)$. Then*

- ***(a)*** *the eigenvalues of $\mathcal{I}_{d_0}$ are $(N-1)/v$ and $(N-1+\alpha_f)/v$ with respective multiplicities $\alpha_f - 1$ and 1;*
- ***(b)*** *d_0 satisfies conditions (i) and (ii) of Lemma 6.2.1.*

Proof.

(a) Let d^* be an $(N-1)$-run plan given by an $OA(N-1, n, m_1 \times \cdots \times m_n, 2f)$ and d_0 be obtained by adding any treatment combination, say $j_1 \ldots j_n$, to d^*. Denote the column of $P^{(1)}$ corresponding to $j_1 \ldots j_n$ by $\boldsymbol{\pi}$. Then by (2.6.8) and (6.1.1),

$$\mathcal{I}_{d_0} = P^{(1)} R_{d^*} (P^{(1)})' + \boldsymbol{\pi}\boldsymbol{\pi}' = \{(N-1)/v\} I_{\alpha_f} + \boldsymbol{\pi}\boldsymbol{\pi}', \qquad (6.2.2)$$

where, as before, I_{α_f} is the identity matrix of order α_f. From (2.3.9) and (2.3.11), proceeding as in the proof of Lemma 2.5.1, each diagonal element of $(P^{(1)})' P^{(1)}$ equals α_f / v. In particular, then $\boldsymbol{\pi}'\boldsymbol{\pi} = \alpha_f / v$. Hence by (6.2.2), part (a) of the lemma follows noting that $\boldsymbol{\pi}$ is an eigenvector corresponding to $(N-1+\alpha_f)/v$ and that vectors orthogonal to $\boldsymbol{\pi}$ constitute the eigenspace of $(N-1)/v$.

(b) From part (a), obviously d_0 satisfies condition (i) of Lemma 6.2.1. Also, by part (a),

$$\begin{aligned}\operatorname{tr}(\mathcal{I}_{d_0}{}^2) &= (\alpha_f - 1)\left\{\frac{N-1}{v}\right\}^2 + \left\{\frac{N-1+\alpha_f}{v}\right\}^2 \\ &= \frac{\alpha_f(N^2 + \alpha_f - 1)}{v^2}. \qquad (6.2.3)\end{aligned}$$

Application of Rao's bound, namely, Theorem 2.6.2, to the $OA(N-1, n, m_1 \times \cdots \times m_n, 2f)$ yields $N - 1 \geq \alpha_f$. Therefore, by (6.1.2) and (6.2.3),

$$\operatorname{tr}(\mathcal{I}_{d_0}{}^2) < \frac{(N\alpha_f)^2}{v^2(\alpha_f - 1)} = \frac{\{\operatorname{tr}(\mathcal{I}_{d_0})\}^2}{\alpha_f - 1},$$

so that d_0 satisfies condition (ii) of Lemma 6.2.1 as well. ■

By Lemma 6.2.2(a), the eigenvalues of $\mathcal{I}_{d_0}$, and hence the performance of d_0 under any generalized criterion of type 1, do not depend on the particular treatment combination that is added to an orthogonal array to get d_0. In consideration of Lemmas 6.2.1 and 6.2.2(b), such a plan d_0 will be optimal in $\mathcal{D}_N$ with respect to every generalized type 1 criterion, provided it satisfies condition (iii) of Lemma 6.2.1. The following result, due to Cheng (1980a), shows that this is the case for 2^n factorials.

Theorem 6.2.1. *With reference to a 2^n factorial, let $d_0(\in \mathcal{D}_N)$ be obtained by adding any treatment combination to a plan given by a symmetric orthogonal array $OA(N-1, n, 2, 2f)$. Then d_0 is a Resolution (f, f) plan which is optimal in $\mathcal{D}_N$ with respect to every generalized criterion of type 1.*

Proof. As noted above, it suffices to check that d_0 satisfies condition (iii) of Lemma 6.2.1. Consider the matrices P_i and P^x as defined in (2.2.2) and (2.2.4), respectively. With a 2^n factorial, namely, with $m_1 = \cdots = m_n = 2$, by (2.2.2), P_i equals

$$\text{either} \quad \left(\tfrac{1}{\sqrt{2}}, \ -\tfrac{1}{\sqrt{2}}\right) \quad \text{or} \quad \left(-\tfrac{1}{\sqrt{2}}, \ \tfrac{1}{\sqrt{2}}\right),$$

for each i. Hence by (2.2.4) and (2.2.5), for each $\boldsymbol{x}$,

$$P^{\boldsymbol{x}} = v^{-\frac{1}{2}} \boldsymbol{p_x}', \tag{6.2.4}$$

where $v = 2^n$ and $\boldsymbol{p_x}$ is a $v \times 1$ vector with elements ± 1. As before, let Ω_f be the set of binary n-tuples with at most f components unity. For any $d \in \mathcal{D}_N$, then by (2.3.9), (6.1.1), and (6.2.4),

$$\begin{aligned}
\text{tr}(\mathcal{I}_d{}^2) &= \text{tr}\{P^{(1)} R_d (P^{(1)})' P^{(1)} R_d (P^{(1)})'\} \\
&= \text{tr}\{R_d (P^{(1)})' P^{(1)} R_d (P^{(1)})' P^{(1)}\} \\
&= \text{tr}\{R_d (\sum_{\boldsymbol{x} \in \Omega_f} (P^{\boldsymbol{x}})' P^{\boldsymbol{x}}) R_d (\sum_{\boldsymbol{y} \in \Omega_f} (P^{\boldsymbol{y}})' P^{\boldsymbol{y}})\} \\
&= \sum \sum_{\boldsymbol{x}, \boldsymbol{y} \in \Omega_f} \text{tr}\{P^{\boldsymbol{x}} R_d (P^{\boldsymbol{y}})' P^{\boldsymbol{y}} R_d (P^{\boldsymbol{x}})'\} \\
&= v^{-2} \sum \sum_{\boldsymbol{x}, \boldsymbol{y} \in \Omega_f} (\boldsymbol{p_x}' R_d \boldsymbol{p_y})^2.
\end{aligned} \tag{6.2.5}$$

Since the elements of $\boldsymbol{p_x}$ are ± 1 for each $\boldsymbol{x}$, recalling the definition of R_d, it follows that for every $\boldsymbol{x}, \boldsymbol{y} \in \Omega_f, \boldsymbol{x} \neq \boldsymbol{y}$, $\boldsymbol{p_x}' R_d \boldsymbol{p_y}$ must equal a linear combination of the quantities $r_d(j_1 \dots j_n)$ with combining coefficients ± 1; namely it must be of the form

$$p_x{}' R_d p_y = \sum\nolimits_1 r_d(j_1 \dots j_n) - \sum\nolimits_2 r_d(j_1 \dots j_n), \qquad (6.2.6)$$

where $\sum_1$ and $\sum_2$ are partial sums over $r_d(j_1 \dots j_n)$ such that for every $j_1 \dots j_n$, $r_d(j_1 \dots j_n)$ is included either under $\sum_1$ or in $\sum_2$ (but not under both). Clearly, as $d \in \mathcal{D}_N$,

$$\sum\nolimits_1 r_d(j_1 \dots j_n) + \sum\nolimits_2 r_d(j_1 \dots j_n) = N. \qquad (6.2.7)$$

Now, by the definition of d_0, an $OA(N-1, n, 2, 2f)$ exists, which implies that N is odd. Since both the terms in the left-hand side of (6.2.7) are integers, it follows that these two terms are unequal and, again invoking the integrality, (6.2.6) yields

$$|p_x{}' R_d p_y| \geq 1, \qquad \text{for each } x, y \in \Omega_f, x \neq y. \qquad (6.2.8)$$

Similarly, for each $x \in \Omega_f$,

$$p_x{}' R_d p_x = \sum r_d(j_1 \dots j_n) = N, \qquad (6.2.9)$$

as $d \in \mathcal{D}_N$ and the elements of p_x are ± 1. In (6.2.9), $\sum$ denotes sum over all possible $j_1 \dots j_n$.

Since $m_1 = \cdots = m_n = 2$, by (2.3.11), the cardinality of Ω_f equals α_f. Hence by (6.2.5), (6.2.8), and (6.2.9), for any $d \in \mathcal{D}_N$,

$$\text{tr}(\mathcal{I}_d{}^2) \geq v^{-2}\{\alpha_f N^2 + \alpha_f(\alpha_f - 1)\} = \frac{\alpha_f(N^2 + \alpha_f - 1)}{v^2}. \qquad (6.2.10)$$

But as seen in (6.2.3), the right-hand side of (6.2.10) equals $\text{tr}(\mathcal{I}_{d_0}{}^2)$. Thus d_0 satisfies condition (iii) of Lemma 6.2.1, and the result follows. ∎

It is of interest to examine how far Theorem 6.2.1 can be extended to general $m_1 \times \cdots \times m_n$ factorials. This appears to be a difficult problem, since equation (6.2.4) and the related fact that the elements of p_x are ± 1, which were crucial in proving Theorem 6.2.1, are no longer valid for a general $m_1 \times \cdots \times m_n$ setup. With reference to a 3^n factorial, Kolyva-Machera (1989a) proved the D-optimality of a plan obtained by the augmentation of any single run to a three-symbol orthogonal array of strength *two*. The proof is via induction on n. Collombier (1988) worked with a complex-valued parametrization and reported an extension of Theorem 6.2.1 to general $m_1 \times \cdots \times m_n$ factorials. One might be inclined to believe that Collombier's (1988) result, obtained via a complex parametrization, can be routinely translated to the practically more meaningful case of real parametrization under which we work here. If so, then this would completely settle the issue of optimality of orthogonal array plus one run plans. Unfortunately, as noted in Mukerjee (1999), there is a gap in Collombier's (1988) proof even under his complex parametrization and this renders such a translation of his findings to the case of a real parametrization impossible. In fact Mukerjee (1999) presented the following example which shows that an extension of Theorem 6.2.1 to $m_1 \times \cdots \times m_n$ factorials is not possible for arbitrary $m_1, \ldots, m_n$:

Example 6.2.1. Let A be an orthogonal array $OA(144, 131, 2^{119} \times 3^{12}, 2)$ which, following Section 4.4, can be constructed as

$$A = [A_1 \odot D_1 \vdots \mathbf{0}_4 \odot A_2],$$

where A_1 is a symmetric $OA(4, 3, 2, 2)$, D_1 is a difference matrix $D(36, 36; 2)$ arising from a Hadamard matrix of order 36 (*vide* Section 3.5), $\mathbf{0}_4 = (0, 0, 0, 0)'$, A_2 is an $OA(36, 23, 2^{11} \times 3^{12}, 2)$ and $\odot$ represents the Kronecker sum. Note that A_1 is given by Theorem 3.2.2, while A_2 can be constructed, again following Section 4.4, as

$$A_2 = [\mathbf{0}_3 \odot A_3 \vdots \boldsymbol{\lambda} \odot D_2],$$

where $\mathbf{0}_3 = (0, 0, 0)'$, A_3 is a symmetric $OA(12, 11, 2, 2)$ available via Theorem 3.2.2, $\boldsymbol{\lambda} = (0, 1, 2)'$ and D_2 is a difference matrix $D(12, 12; 3)$, displayed by Wang and Wu (1991) (see also Remark 3.5.3).

Let d_0 be a plan obtained by adding any run to the array A. Then, with $f = 1$, $N = 145$ and $\alpha_f = 144$, it follows from Lemma 6.2.2(a) that the eigenvalues of $\mathcal{I}_{d_0}$ are $144/v$ and $288/v$, the respective multiplicities being 143 and 1; here $v = (2^{119})(3^{12})$. Hence

$$\operatorname{tr}({\mathcal{I}_{d_0}}^2) = \frac{3048192}{v^2}, \tag{6.2.11}$$

$$\det(\mathcal{I}_{d_0}) = \frac{5971968(144^{141})}{v^{144}}, \tag{6.2.12}$$

$$\operatorname{tr}({\mathcal{I}_{d_0}}^{-1}) = 0.996528v. \tag{6.2.13}$$

Without loss of generality, suppose that the first row of A is $00\ldots0$. Consider a rival plan d_1 obtained from the array A as follows: delete the run $00\ldots0$ given by the first row of A and then add the two runs $11\ldots1$ and $11\ldots122\ldots2$ where, in the latter, the first 119 entries are 1 and the rest are 2. Just as d_0, the plan d_1 involves 145 runs and, with $f = 1$, analogously to (6.2.2),

$$\mathcal{I}_{d_1} = \left(\frac{144}{v}\right) I_{144} - \boldsymbol{\pi}_0\boldsymbol{\pi}_0{}' + \boldsymbol{\pi}_1\boldsymbol{\pi}_1{}' + \boldsymbol{\pi}_2\boldsymbol{\pi}_2{}', \tag{6.2.14}$$

where $\boldsymbol{\pi}_0$, $\boldsymbol{\pi}_1$, and $\boldsymbol{\pi}_2$ are the columns of $P^{(1)}$ corresponding to the treatment combinations $00\ldots0$, $11\ldots1$ and $11\ldots122\ldots2$, respectively. As noted in Mukerjee (1999), by (6.2.14), the eigenvalues of $\mathcal{I}_{d_1}$ are $144/v$, $180/v$ and $(198 \pm 2\sqrt{1351})/v$, of which the first has multiplicity 141 and the others have multiplicity one each. Hence

$$\operatorname{tr}(\mathcal{I}_{d_1}{}^2) = \frac{3045392}{v^2}, \tag{6.2.15}$$

$$\det(\mathcal{I}_{d_1}) = \frac{6084000(144^{141})}{v^{144}}, \tag{6.2.16}$$

$$\operatorname{tr}(\mathcal{I}_{d_1}{}^{-1}) = 0.996438v. \tag{6.2.17}$$

By (6.2.11) and (6.2.15), d_0 does not minimize $\operatorname{tr}(\mathcal{I}_d{}^2)$ over $\mathcal{D}_{145}$ and thus fails to satisfy condition (iii) of Lemma 6.2.1. What is more important, by (6.2.12), (6.2.13), (6.2.16), and (6.2.17), it is dominated by d_1 under both the D- and A-criteria. Thus, for arbitrary $m_1, \ldots, m_n$, an orthogonal array plus one run plan is not necessarily optimal, over the class of plans having the same number of runs, with respect to every generalized criterion of type 1.

■

In view of the above example, which is quite counterintuitive, it is worthwhile to develop conditions on $m_1, \ldots, m_n$ that allow an extension of Theorem 6.2.1. Mukerjee (1999) gave a broad sufficient condition which is as follows:

Theorem 6.2.2. *Suppose that there exists an orthogonal array $OA(N-1, n, m_1 \times \cdots \times m_n, 2f)$, and let $d_0 (\in \mathcal{D}_N)$ be obtained by adding any treatment combination to the plan given by this array. Then d_0 is a Resolution (f, f) plan. Furthermore d_0 is optimal in $\mathcal{D}_N$ with respect to every generalized criterion of type 1 if*

$$\text{H.C.F.}(m_{i_1}, \ldots, m_{i_{2f}}) \geq 2$$
$$\textit{for each } i_1, \ldots, i_{2f} (1 \leq i_1 < \cdots < i_{2f} \leq n), \tag{6.2.18}$$

where H.C.F. *stands for the highest common factor.* ■

From Remark 6.2.1 and Lemma 6.2.2, it is evident that d_0 is a Resolution (f, f) plan. The proof of its optimality, as detailed in Mukerjee (1999), consists of two main steps which are indicated below. In what follows, $W^{\boldsymbol{x}}$ is as defined in (2.5.3) and (2.5.4), and for any $d \in \mathcal{D}_N$, $\boldsymbol{r}_d$ is a $v \times 1$ vector with elements $r_d(j_1 \dots j_n)$ arranged lexicographically.

Step 1. For every $d \in \mathcal{D}_N$,

$$\mathrm{tr}(\mathcal{I}_d{}^2) = \sum_{\boldsymbol{x} \in \Omega_{2f}} a(\boldsymbol{x}) \boldsymbol{r}_d{}' W^{\boldsymbol{x}} \boldsymbol{r}_d, \tag{6.2.19}$$

where the scalars $a(\boldsymbol{x})$ are nonnegative and do not depend on d.

Step 2. If (6.2.18) holds then for every $d \in \mathcal{D}_N$ and every $\boldsymbol{x} \in \Omega_{2f}$,

$$\boldsymbol{r}_d{}' W^{\boldsymbol{x}} \boldsymbol{r}_d \geq \boldsymbol{r}_{d_0}{}' W^{\boldsymbol{x}} \boldsymbol{r}_{d_0}. \tag{6.2.20}$$

From the above, d_0 satisfies condition (iii) of Lemma 6.2.1, and its optimality, as claimed in Theorem 6.2.2, follows. The verification of (6.2.19) and (6.2.20) is rather involved and the interested reader is referred to the original paper of Mukerjee (1999) for details.

Example 6.2.1 shows that the conclusion of Theorem 6.2.2 may not hold without the condition (6.2.18). In fact, in the absence of (6.2.18), the conclusion of Step 1 remains valid while that of Step 2 may break down.

Theorem 6.2.2 has a wide applicability. In particular, (6.2.18) certainly holds for symmetric $m^n (m \geq 2)$ factorials. Thus Theorem 6.2.2 substantially strengthens Theorem 6.2.1, and the findings in Kolyva-Machera (1989a) on 3^n factorials and completely settles the issue of optimality of orthogonal array plus one-run plans for symmetric factorials. It is also applicable to many asymmetric factorials of practical interest, including those where $m_1, \dots, m_n$ are powers of the same integer (e.g., $2^{n_1} \times 4^{n_2}$ factorials). In particular, from Theorem 6.2.2 it is clear that if a single run is added to any of the orthogonal arrays given by Theorems 4.3.1–4.3.4, 4.5.2, 4.5.3, 4.5.5, 4.5.6, 4.6.1, or 4.6.2, then the resulting plan is optimal over the class of plans having the same number of runs with respect to any generalized criterion of type 1.

6.3 AUGMENTED ORTHOGONAL ARRAYS: FURTHER OPTIMALITY RESULTS

In this section we will be primarily concerned with the optimality of plans obtained by adding two or more runs to an orthogonal array. At the outset, we note that, unlike what happens with orthogonal array plus one-run plans, the behavior of plans obtained via augmentation of an orthogonal array by two or more runs often depends on the particular runs added. As the next example illustrates, this can happen even in the simplest case of 2^n factorials and even when only two runs are added.

Example 6.3.1. Let A be a symmetric $OA(8, 7, 2, 2)$ as given by Theorem 3.2.2. Consider two fractions d_0 and d_1 of a 2^7 factorial such that d_0 is obtained by adding the two runs 1111111 and 1110000 to A and d_1 is obtained by adding the two runs 1111111 and 0000000 to A. Both d_0 and d_1 belong to $\mathcal{D}_{10}$, and with $f = 1$, it follows from (6.1.1) that, analogously to (6.2.2),

$$\mathcal{I}_{d_0} = \left(\frac{8}{v}\right) I_8 + \boldsymbol{\pi}_1\boldsymbol{\pi}_1' + \boldsymbol{\pi}_2\boldsymbol{\pi}_2',$$

$$\mathcal{I}_{d_1} = \left(\frac{8}{v}\right) I_8 + \boldsymbol{\pi}_1\boldsymbol{\pi}_1' + \boldsymbol{\pi}_0\boldsymbol{\pi}_0',$$

where $v = 2^7$, and $\boldsymbol{\pi}_0$, $\boldsymbol{\pi}_1$ and $\boldsymbol{\pi}_2$ are columns of $P^{(1)}$ corresponding to the treatment combinations 0000000, 1111111, and 1110000, respectively.

In order to get $\boldsymbol{\pi}_0$, $\boldsymbol{\pi}_1$, and $\boldsymbol{\pi}_2$ explicitly, we choose the matrices P_i, defined in (2.2.2), as $P_i = (-\frac{1}{\sqrt{2}}, \frac{1}{\sqrt{2}})$ for each i. As noted in Remark 2.3.1, this does not entail any loss of generality. Then with

$$P^{(1)} = (P^{00\ldots0'}, P^{10\ldots0'}, P^{01\ldots0'}, \ldots, P^{00\ldots1'})', \tag{6.3.1}$$

by (2.2.4) and (2.2.5), it is seen that

$$\boldsymbol{\pi}_0 = v^{-1/2}(1, -1, -1, -1, -1, -1, -1, -1)',$$

$$\boldsymbol{\pi}_1 = v^{-1/2}(1, 1, 1, 1, 1, 1, 1, 1)',$$

$$\boldsymbol{\pi}_2 = v^{-1/2}(1, 1, 1, 1, -1, -1, -1, -1)'.$$

Hence the eigenvalues of $\mathcal{I}_{d_0}$ are $16/v$ and $8/v$ with respective multiplicities 2 and 6, while those of $\mathcal{I}_{d_1}$ are $22/v$, $10/v$ and $8/v$ with respective multiplicities 1, 1, 6. This yields

$$\det(\mathcal{I}_{d_0}) = \frac{67108864}{v^8}, \quad \operatorname{tr}(\mathcal{I}_{d_0}{}^{-1}) = 0.875v,$$

$$\det(\mathcal{I}_{d_1}) = \frac{57671680}{v^8}, \quad \operatorname{tr}(\mathcal{I}_{d_1}{}^{-1}) = 0.895v,$$

showing the superiority of d_0 over d_1 under both the D- and A-criteria despite the fact that both of these plans are obtained by adding two runs to the same orthogonal array. Incidentally, the smallest eigenvalues of $\mathcal{I}_{d_0}$ and $\mathcal{I}_{d_1}$ are identical, so that the two plans are equivalent under the E-criterion. ■

As Example 6.3.1 illustrates, under commonly used criteria, like the D- or A-criteria, plans obtained via the addition of two or more runs to an orthogonal array are not all equivalent. Thus it is important to identify the best way of augmentation, under a given criterion, while attempting to study the optimality of such plans. Example 6.3.1 also suggests that the E-criterion might be an exception in the sense that augmented orthogonal arrays tend to be equivalent under this criterion at least when the number of runs added is not too large. This point will be made more explicit later in Theorems 6.3.3–6.3.6.

While proving Theorem 6.2.1, in (6.2.4), we noted that for 2^n factorials the matrices P^x reduce to the form $P^x = v^{-1/2}\boldsymbol{p}_x{}'$, the elements of the vectors $\boldsymbol{p}_x$ being ± 1. Readers familiar with chemical balance weighing designs (i.e., designs where the elements of the design matrix are 0 or ± 1; see Banerjee, 1975) might therefore anticipate a link between optimal weighing designs and optimal fractions of 2^n factorials. Indeed, as can be seen from Cheng (1980a), such a link exists in the case of Theorem 6.2.1. We now present another result in this direction. Some preliminary results, including one on majorization, are needed for this purpose.

Consider $u \times 1$ vectors $\boldsymbol{c}^1 = (c_1{}^1, \ldots, c_u{}^1)'$ and $\boldsymbol{c}^2 = (c_1{}^2, \ldots, c_u{}^2)'$, and let $c_{[1]}{}^1 \geq c_{[2]}{}^1 \geq \cdots \geq c_{[u]}{}^1$ and $c_{[1]}{}^2 \geq c_{[2]}{}^2 \geq \cdots \geq c_{[u]}{}^2$ be the ordered elements of $\boldsymbol{c}^1$ and $\boldsymbol{c}^2$, respectively. Then $\boldsymbol{c}^1$ is said to be majorized by $\boldsymbol{c}^2$ if

$$\sum_{i=1}^{j} c_{[i]}{}^1 \leq \sum_{i=1}^{j} c_{[i]}{}^2, \qquad 1 \leq j \leq u-1$$

and

$$\sum_{i=1}^{u} c_{[i]}{}^1 = \sum_{i=1}^{u} c_{[i]}{}^2.$$

For example, if $\boldsymbol{c}^1 = (1/u, 1/u, \ldots, 1/u)'$ and $\boldsymbol{c}^2 = (1, 0, \ldots, 0)'$, then $\boldsymbol{c}^1$ is majorized by $\boldsymbol{c}^2$.

A square matrix with nonnegative elements is called a doubly stochastic matrix if each of its row and column sums is equal to unity. We are now in a position to state the following useful lemma, whose proof can be found in Marshall and Olkin (1979, pp. 22 and 225).

Lemma 6.3.1.

(a) *A necessary and sufficient condition for* $\mathbf{c}^1$ *to be majorized by* $\mathbf{c}^2$ *is that* $\mathbf{c}^1 = \Gamma \mathbf{c}^2$ *for some doubly stochastic matrix* Γ.

(b) *Let* Λ *and* $\bar{\Lambda}$ *be* $u \times u$ *symmetric matrices of the form*

$$\Lambda = \begin{pmatrix} \Lambda_{11} & \Lambda_{12} \\ \Lambda_{21} & \Lambda_{22} \end{pmatrix}, \quad \bar{\Lambda} = \begin{pmatrix} \Lambda_{11} & O \\ O & \Lambda_{22} \end{pmatrix}$$

and $\mathbf{c}^1$ *and* $\mathbf{c}^2$ *be* $u \times 1$ *vectors with elements given by the eigenvalues of* $\bar{\Lambda}$ *and* Λ, *respectively. Then* $\mathbf{c}^1$ *is majorized by* $\mathbf{c}^2$. ■

Lemma 6.3.2. *Let* $u \equiv 2 \pmod 4$ *and* $\boldsymbol{p}_1$, $\boldsymbol{p}_2$ *be* $u \times 1$ *vectors with elements* ± 1 *such that* $\boldsymbol{p}_i$ *has* u_i *elements equal to* 1 $(i = 1, 2)$. *If* u_1 *and* u_2 *are both even or both odd, then* $|\boldsymbol{p}_1' \boldsymbol{p}_2| \geq 2$.

Proof. This lemma, due to Jacroux et al. (1983), follows by noting that if u_0 be the number of positions where both $\boldsymbol{p}_1$ and $\boldsymbol{p}_2$ have 1, then

$$\boldsymbol{p}_1' \boldsymbol{p}_2 = 2\left(\frac{u}{2} - u_1 - u_2 + 2u_0\right). \tag{6.3.2}$$

Now $u/2$ is an odd integer as $u \equiv 2 \pmod 4$. Hence, if u_1 and u_2 are both even or both odd, then the integer $u/2 - u_1 - u_2 + 2u_0$ is nonzero. Therefore the result follows from (6.3.2). ■

We also need the following lemma on the function q introduced in (6.2.1) while defining an optimality criterion of type 1. The proof of this lemma is available in Cheng (1978).

Lemma 6.3.3. *Let* q *be a real-valued function defined on* $[0, N\alpha_f/v]$ *and satisfying the conditions (a) and (b) stated in connection with (6.2.1). Furthermore, assume that* $\lim_{\mu \to 0^+} q(\mu) = q(0) = +\infty$. *Let* $u (\geq 2)$ *be a given integer, and let* κ_1 *and* κ_2 *be given positive quantities such that* ${\kappa_1}^2 \geq \kappa_2 \geq {\kappa_1}^2/u$ *and* $\kappa_1 \leq N\alpha_f/v$. *Define*

$$\omega_1 = \frac{\kappa_1 + \{(u-1)(u\kappa_2 - {\kappa_1}^2)\}^{1/2}}{u},$$

$$\omega_2 = \frac{\kappa_1 - \{(u\kappa_2 - {\kappa_1}^2)/(u-1)\}^{1/2}}{u}.$$

Then $0 \le \omega_2 \le \omega_1 \le N\alpha_f/v$, *and the minimum of* $\sum_{i=1}^{u} q(\mu_i)$ *over the set*

$$\left\{(\mu_1, \ldots, \mu_u)' : \mu_i \ge 0 \text{ for all } i,\ \sum_{i=1}^{u} \mu_i = \kappa_1,\ \sum_{i=1}^{u} {\mu_i}^2 = \kappa_2\right\}$$

equals $q(\omega_1) + (u-1)q(\omega_2)$. ■

We now have the following result obtained originally by Jacroux et al. (1983) in the context of weighing designs.

Theorem 6.3.1. *Consider a* 2^n *factorial. Let* $l_1 \ldots l_n$ *and* $l_1{}^* \ldots l_n{}^*$ *be any two treatment combinations such that* $l_i = l_i{}^*$ *for exactly* $\left[\frac{1}{2}n\right]$ *choices of* i, *where* $\left[\frac{1}{2}n\right]$ *is the largest integer in* $\frac{1}{2}n$. *Let* $d_0 (\in \mathcal{D}_N)$ *be obtained by adding two such treatment combinations to a plan given by a symmetric* $OA(N-2, n, 2, 2)$. *Then* d_0 *is a Resolution* $(1, 1)$ *plan that is optimal in* $\mathcal{D}_N$ *with respect to every type 1 criterion, as defined in (6.2.1), for which* $\lim_{\mu \to 0^+} q(\mu) = q(0) = +\infty$.

Proof. By the definition of d_0, an $OA(N-2, n, 2, 2)$ exists, which implies that

$$N - 2 \ge 4 \quad \text{and} \quad N \equiv 2 \bmod 4. \tag{6.3.3}$$

Here $f = 1$ and, by (2.3.11), $\alpha_f = n + 1$. Hence by (2.3.9), $P^{(1)}$ is of order $(n+1) \times v$, where $v = 2^n$. Let $\pi(j_1 \ldots j_n)$ be the column of $P^{(1)}$ corresponding to any treatment combination $j_1 \ldots j_n$. For any $d \in \mathcal{D}_N$, as in Section 5.4, let G_d be an $(n+1) \times N$ matrix such that for $1 \le i \le N$, if the ith run in d is given by the treatment combination $u_1 \ldots u_n$, then the ith column of G_d equals $\pi(u_1 \ldots u_n)$. Then from (6.1.1) it is easily seen that

$$\mathcal{I}_d = G_d G_d{}'. \tag{6.3.4}$$

Since a 2^n factorial is being considered, by (6.2.4), each element of G_d is $\pm v^{-1/2}$, namely

$$G_d = v^{-1/2} K_d, \tag{6.3.5}$$

where the $(n+1) \times N$ matrix K_d has entries ± 1. Let n_1 be the number of rows of K_d having an even number of 1's, K_{d1} be an $n_1 \times N$ submatrix of K_d consisting of these n_1 rows, and K_{d2} be an $n_2 \times N$ submatrix of K_d consisting of the remaining $n_2 = n + 1 - n_1$ rows, each with an odd number of 1's.

First suppose that $n_1 \geq 2$ and $n_2 \geq 2$. By (6.3.4) and (6.3.5) then, up to a permutation of rows and columns,

$$\mathcal{I}_d = v^{-1}\begin{pmatrix} K_{d1}K_{d1}' & K_{d1}K_{d2}' \\ K_{d2}K_{d1}' & K_{d2}K_{d2}' \end{pmatrix}.$$

Denote the eigenvalues of $\mathcal{I}_d$, $v^{-1}K_{d1}K_{d1}'$ and $v^{-1}K_{d2}K_{d2}'$ by $\mu_{di}(1 \leq i \leq n+1)$, $\mu_{di}{}^{(1)}(1 \leq i \leq n_1)$ and $\mu_{di}{}^{(2)}(1 \leq i \leq n_2)$ respectively. Then by Lemma 6.3.1(b), the vector $(\mu_{d1}{}^{(1)}, \ldots, \mu_{dn_1}{}^{(1)}, \mu_{d1}{}^{(2)}, \ldots \mu_{dn_2}{}^{(2)})'$ is majorized by the vector with elements $\mu_{di}(1 \leq i \leq n+1)$. Hence using Lemma 6.3.1(a) and recalling the convexity of q, we have

$$\sum_{i=1}^{n+1} q(\mu_{di}) \geq \sum_{i=1}^{n_1} q(\mu_{di}{}^{(1)}) + \sum_{i=1}^{n_2} q(\mu_{di}{}^{(2)}). \tag{6.3.6}$$

Since the elements of K_d are ± 1, each diagonal element of $K_{d1}K_{d1}'$ is N. Hence

$$\sum_{i=1}^{n_1} \mu_{di}{}^{(1)} = v^{-1}\text{tr}(K_{d1}K_{d1}') = \frac{n_1 N}{v}. \tag{6.3.7}$$

Also, since each row of K_{d1} is a $1 \times N$ vector with an even number of 1's, by (6.3.3) and Lemma 6.3.2, it follows that the absolute value of every off-diagonal element of $K_{d1}K_{d1}'$ is at least 2. Therefore

$$\sum_{i=1}^{n_1} \{\mu_{di}{}^{(1)}\}^2 = v^{-2}\text{tr}\{(K_{d1}K_{d1}')^2\} \geq v^{-2}\{n_1 N^2 + 4n_1(n_1 - 1)\}. \tag{6.3.8}$$

In view of (6.3.7), (6.3.8), and the convexity of q, an application of Lemma 6.3.3 yields

$$\sum_{i=1}^{n_1} q(\mu_{di}{}^{(1)}) \geq q\left(\frac{N + 2(n_1 - 1)}{v}\right) + (n_1 - 1)q\left(\frac{N-2}{v}\right).$$

Using a similar inequality for $\sum_{i=1}^{n_2} q(\mu_{di}{}^{(2)})$ and recalling that $n_1 + n_2 = n + 1$, from (6.3.6) one obtains

$$\sum_{i=1}^{n+1} q(\mu_{di}) \geq q\left(\frac{N + 2(n_1 - 1)}{v}\right) + q\left(\frac{N + 2(n - n_1)}{v}\right) + (n-1)q\left(\frac{N-2}{v}\right). \tag{6.3.9}$$

In a similar manner one can verify that (6.3.9) holds also when $n_1 = 0$ or 1 or $n_2 = 0$ or 1.

Next observe that

$$\left\{\frac{N+2(n_1-1)}{v}\right\} + \left\{\frac{N+2(n-n_1)}{v}\right\} = \frac{2(N+n-1)}{v},$$

which does not depend on n_1. Hence, using the convexity of q, the sum of the first two terms in the right-hand side of (6.3.9) is minimized with respect to the integer n_1 when $n_1 = \left[\frac{1}{2}n\right]+1$. Consequently, by (6.3.9), for any $d \in \mathcal{D}_N$,

$$\sum_{i=1}^{n+1} q(\mu_{di}) \geq q\left(\frac{N+2[n/2]}{v}\right) + q\left(\frac{N-2+2(n-[n/2])}{v}\right) + (n-1)q\left(\frac{N-2}{v}\right). \qquad (6.3.10)$$

Consider now the plan d_0. Analogously to (6.2.2) or, Example 6.3.1,

$$\mathcal{I}_{d_0} = \left\{\frac{N-2}{v}\right\} I_{n+1} + \boldsymbol{\pi}_1\boldsymbol{\pi}_1' + \boldsymbol{\pi}_2\boldsymbol{\pi}_2', \qquad (6.3.11)$$

where $\boldsymbol{\pi}_1$ and $\boldsymbol{\pi}_2$ are columns of $P^{(1)}$ corresponding to the two treatment combinations that are added to an $OA(N-2, n, 2, 2)$ to get d_0. Without loss of generality, suppose that exactly the first $\left[\frac{1}{2}n\right]$ levels are identical in these two treatment combinations. Then, as in Example 6.3.1, with $P^{(1)}$ given by (6.3.1),

$$\boldsymbol{\pi}_j = v^{-1/2}(1, a_{1j}, \ldots, a_{nj})', \qquad j = 1, 2,$$

where $a_{ij} = \pm 1$ for every i, j, and $a_{i1} = a_{i2}$ for $1 \leq i \leq \left[\frac{1}{2}n\right]$ while $a_{i1} = -a_{i2}$ for $\left[\frac{1}{2}n\right]+1 \leq i \leq n$. Writing $t_1 = \left[\frac{1}{2}n\right]+1$ and $t_2 = n - \left[\frac{1}{2}n\right]$, then

$$\boldsymbol{\pi}_1 = v^{-1/2}\left(\boldsymbol{a}^{(1)\prime}, \boldsymbol{a}^{(2)\prime}\right)',$$

$$\boldsymbol{\pi}_2 = v^{-1/2}\left(\boldsymbol{a}^{(1)\prime}, -\boldsymbol{a}^{(2)\prime}\right)', \qquad (6.3.12)$$

where $\boldsymbol{a}^{(1)}$ and $\boldsymbol{a}^{(2)}$ are vectors of orders $t_1 \times 1$ and $t_2 \times 1$, respectively, and having elements ± 1. By (6.3.11) and (6.3.12),

$$\mathcal{I}_{d_0} = v^{-1}\begin{pmatrix} (N-2)I_{t_1} + 2\boldsymbol{a}^{(1)}\boldsymbol{a}^{(1)\prime} & O \\ O & (N-2)I_{t_2} + 2\boldsymbol{a}^{(2)}\boldsymbol{a}^{(2)\prime} \end{pmatrix}.$$

Since the elements of $\boldsymbol{a}^{(1)}$ and $\boldsymbol{a}^{(2)}$ are ± 1, it follows that the eigenvalues of $\mathcal{I}_{d_0}$ are $(N - 2 + 2t_1)/v$, $(N - 2 + 2t_2)/v$ and $(N - 2)/v$ with respective multiplicities 1, 1 and $n - 1$. By (6.3.3), these eigenvalues are all positive, so d_0 is a Resolution (1, 1) plan (see Theorem 2.3.1). Furthermore, recalling the definitions of t_1 and t_2, the right-hand side of (6.3.10) is seen to equal $\sum_{i=1}^{n+1} q(\mu_{d_0 i})$, whence the claimed optimality of d_0 follows. ■

Remark 6.3.1. As detailed in Cheng (1978), the proof of Lemma 6.3.3 involves application of the method of Lagrangian multipliers, and this requires the condition $\lim_{\mu \to 0^+} q(\mu) = q(0) = +\infty$. Consequently the same condition is needed in Theorem 6.3.1 as well. It is worth noting that the class of type 1 criteria satisfying this condition is quite rich and includes the D- and A-criteria in particular. Clearly, by Theorem 6.3.1, the plan d_0 considered in Example 6.3.1 is optimal in $\mathcal{D}_{10}$ with respect to every such type 1 criterion.

■

Chadjiconstantinidis et al. (1989) obtained the following extension of Theorem 6.3.1. The proof resembles that of Theorem 6.3.1 but requires more intricate arguments.

Theorem 6.3.2. *Consider a 2^n factorial. Let $l_1 \ldots l_n$ and $l_1{}^* \ldots l_n{}^*$ be any two treatment combinations such that $l_i = l_i{}^*$ for exactly u_0 choices of i, where u_0 maximizes*

$$\min\left\{1 + \binom{n}{2} - u(n-u-1),\, n + u(n-u-1)\right\}$$

over $u \in \{0, 1, \ldots, n\}$. Let $d_0 (\in \mathcal{D}_N)$ be obtained by adding two such treatment combinations to a plan given by a symmetric $OA(N-2, n, 2, 4)$. Then d_0 is a Resolution (2, 2) *plan that is optimal in $\mathcal{D}_N$ with respect to every type 1 criterion, as defined in (6.2.1), for which* $\lim_{\mu \to 0^+} q(\mu) = q(0) = +\infty$.

■

Example 6.3.2. Let A be a symmetric $OA(16, 5, 2, 4)$ as given by Theorem 3.4.2. Then $n = 5$ and the quantity u_0, defined in Theorem 6.3.2, equals 1 or 3. Take $u_0 = 1$, and consider the plan d_0 obtained by adding the runs 11111 and 10000 to A. Then d_0 is a Resolution (2, 2) plan, which is optimal in $\mathcal{D}_{18}$ in the sense of Theorem 6.3.2. ■

Turning to the E-criterion, we have the following result due to Cheng (1980a). From Section 2.5 recall that the E-criterion seeks to minimize $\mu_{\max}(\mathcal{I}_d{}^{-1})$ while keeping $\mathcal{I}_d$ positive definite, or equivalently, attempts to maximize $\mu_{\min}(\mathcal{I}_d)$, where $\mu_{\max}(\cdot)$ and $\mu_{\min}(\cdot)$ stand, respectively, for the maximum and minimum eigenvalues.

Theorem 6.3.3. *Consider a 2^n factorial, where $n \geq 3$. Let $1 \leq u \leq 3$ and $d_0 (\in \mathcal{D}_{N_0+u})$ be obtained by adding any u treatment combinations to a plan given by a symmetric $OA(N_0, n, 2, 2f)$. Then d_0 is a Resolution (f, f) plan that is E-optimal in $\mathcal{D}_{N_0+u}$.*

Proof. From (6.1.1), it is easily seen that analogously to (6.2.2),

$$\mathcal{I}_{d_0} = \left(\frac{N_0}{v}\right) I_{\alpha_f} + P^{(1)} R_{d^*} (P^{(1)})',$$

where $v = 2^n$ and d^* is a subdesign of d_0 consisting of the u runs that are added to an orthogonal array to get d_0. Hence $\mathcal{I}_{d_0} - (N_0/v) I_{\alpha_f}$ is nonnegative definite (n.n.d.) and

$$\mu_{\min}(\mathcal{I}_{d_0}) \geq \frac{N_0}{v}. \tag{6.3.13}$$

Therefore, to complete the proof, it suffices to show that for each $d \in \mathcal{D}_{N_0+u}$,

$$\mu_{\min}(\mathcal{I}_d) \leq \frac{N_0}{v}. \tag{6.3.14}$$

Since $u \leq 3$, any $d' \in \mathcal{D}_{N_0+u}$ is embedded in some $d \in \mathcal{D}_{N_0+3}$ (i.e., d' can be obtained from d through possible deletion of some runs of the latter), so by (6.1.1), $\mathcal{I}_d - \mathcal{I}_{d'}$ is n.n.d. and $\mu_{\min}(\mathcal{I}_{d'}) \leq \mu_{\min}(\mathcal{I}_d)$. Therefore it will suffice to prove (6.3.14) for each $d \in \mathcal{D}_{N_0+3}$.

To that effect, write $T = N_0 + 3$ and consider any $d \in \mathcal{D}_T$. Since a 2^n factorial is being considered, exactly as in the proof of Theorem 6.3.1 (see (6.3.4) and (6.3.5)),

$$\mathcal{I}_d = v^{-1} K K', \tag{6.3.15}$$

where K is of order $\alpha_f \times T$ with elements ± 1; we drop the subscript d of K for notational simplicity. By the definition of d_0, an $OA(N_0, n, 2, 2f)$ exists which, in particular, implies that N_0 is an integral multiple of 4, namely, $T = 3 \bmod 4$. Since T is odd, each row of K contains either an even number of 1's and an odd number of -1's or an odd number of 1's and an even number of -1's. But the eigenvalues of KK' and hence those of $\mathcal{I}_d$ remain unaltered if any row of K is multiplied by -1. Hence, without loss of generality, suppose that each row of K has an even number of 1's. Since $T = 3 \bmod 4$, proceeding as in the proof of Lemma 6.3.2, it can then be seen that the scalar product of every two distinct rows of K, or equivalently, every off-diagonal element of KK' equals 3 mod 4. Also each diagonal element of KK' equals T as the elements of K are ± 1.

If each off-diagonal element of KK' equals -1, then it is easily seen that

$$\mu_{\min}(KK') = T - (\alpha_f - 1) = N_0 + 4 - \alpha_f \leq N_0,$$

since $\alpha_f \geq \alpha_1 = n + 1 \geq 4$, and the truth of (6.3.14) follows from (6.3.15). On the other hand, if any off-diagonal element of KK' is different from -1, then, in consideration of what has been mentioned in the last paragraph, the absolute value of such an element must be at least 3; that is, KK' must have a 2×2 principal submatrix of the form

$$\begin{pmatrix} T & a \\ a & T \end{pmatrix},$$

where $|a| \geq 3$. Since the minimum eigenvalue of this submatrix cannot obviously exceed $T - 3 (= N_0)$, one gets $\mu_{\min}(KK') \leq N_0$, whence the truth of (6.3.14) follows again using (6.3.15). ■

Remark 6.3.2. As noted in Cheng (1980a), Theorem 6.3.3 is linked with weighing designs. By Theorem 6.3.3, the plan d_0 considered in Theorem 6.3.2 is E-optimal within the relevant class. Also, for $n \geq 3$, so is the plan d_0 of Theorem 6.3.1. Some comments on the condition $n \geq 3$, appearing in the statement of Theorem 6.3.3, are in order. This theorem presumes the existence of an $OA(N_0, n, 2, 2f)$ so that $n \geq 2f$. Therefore, if $f \geq 2$, then the condition $n \geq 3$ is automatically satisfied. On the other hand, for $f = 1$ and $n = 2$ examples can be given to demonstrate that the conclusion of Theorem 6.3.3 is not generally true. This, however, is of little concern since, if $f = 1$ and $n = 2$, then $N_0 \geq 2^2 = v$; that is, $N_0 + u > v$, so the designs in $\mathcal{D}_{N_0+u}$ are no longer fractional factorial plans. ■

Any generalization of Theorem 6.3.1 or 6.3.2 to $m_1 \times \cdots \times m_n$ or even symmetric $m^n (m \geq 3)$ factorials is as yet unknown. Under the E-criterion, however, Mukerjee (1995) reported some such extensions of Theorem 6.3.3. The proofs in Mukerjee (1995) are rather involved, and in the next theorem we present a proof of one of his results in a special case to give an idea of the underlying approach.

Theorem 6.3.4. *Consider an m^n factorial, where $m \geq 3$. Let $1 \leq u \leq m+1$ and $d_0 (\in \mathcal{D}_{N_0+u})$ be obtained by adding any u treatment combinations to a* *plan given by a symmetric* $OA(N_0, n, m, 2)$. *Then* d_0 *is a Resolution* (1, 1) *plan that is E-optimal in* $\mathcal{D}_{N_0+u}$.

Proof. As in the proof of Theorem 6.3.3, $\mu_{\min}(\mathcal{I}_{d_0}) \geq N_0/v$, where $v = m^n$, and using the same embedding argument as there, it is enough to show that $\mu_{\min}(\mathcal{I}_d) \leq N_0/v$ for every $d \in \mathcal{D}_T$, where $T = N_0+m+1$. By the definition of d_0, an $OA(N_0, n, m, 2)$ exists and hence N_0/m^2 is an integer. Thus

$$N_0 = \theta m^2 \quad \text{and} \quad T = \theta m^2 + m + 1, \tag{6.3.16}$$

for some integer θ. If possible, let there exist a plan $d \in \mathcal{D}_T$ such that

$$\mu_{\min}(\mathcal{I}_d) > \frac{N_0}{v} = \frac{\theta m^2}{v}. \tag{6.3.17}$$

Here $f = 1$ and $P^{(1)}$ is of the form (6.3.1). Hence writing

$$U = (P^{00\ldots0'}, P^{10\ldots0'}, P^{01\ldots0'})', \tag{6.3.18}$$

by (6.1.1), $\mathcal{I}_d$ contains UR_dU' as a principal submatrix. Therefore, by (6.3.17),

$$\mu_{\min}(UR_dU') > \frac{\theta m^2}{v}. \tag{6.3.19}$$

But by (2.2.4), (2.2.5), and (6.3.18),

$$U = v^{-1/2} P_0 \begin{bmatrix} \mathbf{1}'_v \\ I_m \otimes \mathbf{1}'_m \otimes \cdots \otimes \mathbf{1}'_m \\ \mathbf{1}'_m \otimes I_m \otimes \cdots \otimes \mathbf{1}'_m \end{bmatrix}, \tag{6.3.20}$$

where $\mathbf{1}_u$ is, as usual, a $u \times 1$ vector with all elements unity and,

$$P_0 = \operatorname{diag}(1, m^{1/2} P_1, m^{1/2} P_2), \tag{6.3.21}$$

the $(m-1) \times m$ matrices P_1 and P_2 being as defined in (2.2.2). Since $d \in \mathcal{D}_T$, from (6.3.20) one can check that

$$UR_dU' = v^{-1} P_0 \begin{bmatrix} T & \boldsymbol{r}_{1d}' & \boldsymbol{r}_{2d}' \\ \boldsymbol{r}_{1d} & R_{1d} & B_d \\ \boldsymbol{r}_{2d} & B_d' & R_{2d} \end{bmatrix} P_0', \tag{6.3.22}$$

where for $i = 1, 2,$

$$\boldsymbol{r}_{id} = (r_{id}(0), r_{id}(1), \ldots, r_{id}(m-1))', \tag{6.3.23}$$

$$R_{id} = \mathrm{diag}(r_{id}(0), r_{id}(1), \ldots, r_{id}(m-1)), \tag{6.3.24}$$

with $r_{id}(j)$ representing the frequency of occurrence of level j of the ith factor in the plan d, and B_d is an $m \times m$ matrix with (j_1, j_2)th element representing the number of times the level combination $j_1 j_2$ of the first two factors appears in the plan d $(0 \le j_1, j_2 \le m-1)$. Clearly,

$$\boldsymbol{r}_{1d}'\mathbf{1}_m = \boldsymbol{r}_{2d}'\mathbf{1}_m = T \tag{6.3.25}$$

and

$$B_d\mathbf{1}_m = \boldsymbol{r}_{1d}, \qquad B_d'\mathbf{1}_m = \boldsymbol{r}_{2d}. \tag{6.3.26}$$

Observe that by (6.3.23)–(6.3.25),

$$v^{-1}\begin{pmatrix} m^{-1/2}\mathbf{1}'_m \\ P_1 \end{pmatrix} (mR_{1d}) \begin{pmatrix} m^{-1/2}\mathbf{1}_m & P_1' \end{pmatrix}$$

$$= v^{-1}\begin{pmatrix} 1 & \mathbf{0}' \\ \mathbf{0} & m^{1/2}P_1 \end{pmatrix} \begin{pmatrix} T & \boldsymbol{r}_{1d}' \\ \boldsymbol{r}_{1d} & R_{1d} \end{pmatrix} \begin{pmatrix} 1 & \mathbf{0}' \\ \mathbf{0} & m^{1/2}P_1' \end{pmatrix}. \tag{6.3.27}$$

It is clear from (6.3.21) and (6.3.22) that the matrix in the right-hand side of (6.3.27), and hence that in the left-hand side of (6.3.27), is a principal submatrix of UR_dU'. Since by (2.2.2), the matrix $(m^{-1/2}\mathbf{1}_m, P_1')$ is orthogonal, it now follows from (6.3.19) and (6.3.27) that $\mu_{\min}(v^{-1}mR_{1d}) > \theta m^2/v$; that is, by (6.3.24), $r_{1d}(j) \ge \theta m + 1$ for each $j\,(0 \le j \le m-1)$. Recall that the quantities $r_{1d}(j)$ are integers. At the same time, by (6.3.16) and (6.3.25), $\sum_{j=0}^{m-1} r_{1d}(j) = \theta m^2 + m + 1$. Therefore, among $r_{1d}(j)$, $0 \le j \le m-1$, exactly one equals $\theta m + 2$, and the rest are equal to $\theta m + 1$. Without loss of generality, let

$$r_{1d}(0) = \theta m + 2, \quad r_{1d}(j) = \theta m + 1 \qquad (1 \le j \le m-1). \tag{6.3.28}$$

In a similar manner, without loss of generality, we have

$$r_{2d}(0) = \theta m + 2, \quad r_{2d}(j) = \theta m + 1 \qquad (1 \le j \le m-1). \tag{6.3.29}$$

Let $\boldsymbol{e}_0, \boldsymbol{e}_1, \ldots, \boldsymbol{e}_{m-1}$ be the unit vectors of order m and, for any $0 \le j_1, j_2 \le m-1$, $\boldsymbol{\xi} = (0, \boldsymbol{e}_{j_1}'P_1', -\boldsymbol{e}_{j_2}'P_2')'$. Since by (2.2.2), $P_i'P_i = I_m -$

$m^{-1}\mathbf{1}_m\mathbf{1}'_m$ $(i = 1, 2)$, then $\boldsymbol{\xi}'\boldsymbol{\xi} = 2(1 - m^{-1})$, and using (6.3.21)–(6.3.26),

$$\boldsymbol{\xi}'UR_dU'\boldsymbol{\xi} = v^{-1}m\{r_{1d}(j_1) + r_{2d}(j_2) - 2b_d(j_1, j_2)\},$$

where $b_d(j_1, j_2)$ is the (j_1, j_2)th element of B_d. In view of (6.3.19),

$$\frac{\boldsymbol{\xi}'UR_dU'\boldsymbol{\xi}}{\boldsymbol{\xi}'\boldsymbol{\xi}} > \frac{\theta m^2}{v},$$

and this yields

$$r_{1d}(j_1) + r_{2d}(j_2) - 2b_d(j_1, j_2) \geq 2\theta(m - 1) + 1, \tag{6.3.30}$$

noting that the quantities in the left-hand side of (6.3.30) are integers. Since in particular, the quantities $b_d(j_1, j_2)$ are integers, from (6.3.28)–(6.3.30) one obtains

$$b_d(j_1, 0) \leq \theta + 1 \qquad (1 \leq j_1 \leq m - 1), \tag{6.3.31a}$$

$$b_d(0, j_2) \leq \theta + 1 \qquad (1 \leq j_2 \leq m - 1), \tag{6.3.31b}$$

$$b_d(j_1, j_2) \leq \theta \qquad (1 \leq j_1, j_2 \leq m - 1). \tag{6.3.32}$$

By (6.3.26), (6.3.31a,b), and (6.3.32),

$$r_{1d}(j_1) = \sum_{j_2=0}^{m-1} b_d(j_1, j_2) \leq \theta m + 1 \qquad (1 \leq j_1 \leq m - 1),$$

$$r_{2d}(j_2) = \sum_{j_1=0}^{m-1} b_d(j_1, j_2) \leq \theta m + 1 \qquad (1 \leq j_2 \leq m - 1),$$

and comparing the above with (6.3.28), (6.3.29), it follows that equality must hold in (6.3.31a,b), and (6.3.32), namely

$$b_d(j_1, 0) = \theta + 1 \qquad (1 \leq j_1 \leq m - 1), \tag{6.3.33a}$$

$$b_d(0, j_2) = \theta + 1 \qquad (1 \leq j_2 \leq m - 1), \tag{6.3.33b}$$

$$b_d(j_1, j_2) = \theta \qquad (1 \leq j_1, j_2 \leq m - 1). \tag{6.3.34}$$

From (6.3.16), (6.3.25), and (6.3.26), $\sum\sum_{j_1, j_2=0}^{m-1} b_d(j_1, j_2) = \theta m^2 + m + 1$. Hence by (6.3.33a,b) and (6.3.34)

$$b_d(0, 0) = \theta + 3 - m. \tag{6.3.35}$$

Since $b_d(0, 0) \geq 0$, the impossibility of (6.3.17) follows from (6.3.35) if $\theta + 3 < m$.

Continuing with the case $\theta + 3 \geq m$, by (6.3.23), (6.3.24), (6.3.28), (6.3.29), and (6.3.33a,b)–(6.3.35), note that

$$\begin{aligned}
\boldsymbol{r}_{1d} &= \boldsymbol{r}_{2d} = (\theta m + 1)\mathbf{1}_m + \boldsymbol{e}_0, \\
R_{1d} &= R_{2d} = (\theta m + 1)I_m + \boldsymbol{e}_0\boldsymbol{e}_0', \\
B_d &= \theta \mathbf{1}_m \mathbf{1}'_m + \boldsymbol{e}_0 \mathbf{1}'_m + \mathbf{1}_m \boldsymbol{e}_0' - (m-1)\boldsymbol{e}_0\boldsymbol{e}_0'.
\end{aligned}$$

Then, with $\boldsymbol{\xi}_0 = (-m^{-1/2}, \boldsymbol{e}_0'P_1', \boldsymbol{e}_0'P_2')'$, from (6.3.21) and (6.3.22) after some algebra one obtains

$$\frac{\boldsymbol{\xi}_0' U R_d U' \boldsymbol{\xi}_0}{\boldsymbol{\xi}_0'\boldsymbol{\xi}_0} = \frac{\theta m^2(2m-1) - (m-3)\{2(m-1)^2 + 1\}}{(2m-1)v}.$$

By (6.3.19), the right-hand side of the above must exceed $\theta m^2/v$, which is clearly impossible as $m \geq 3$. Thus we again reach a contradiction which completes the proof. ∎

Mukerjee (1995) in fact proved the following more general result which covers both Theorems 6.3.3 and 6.3.4 as special cases and strengthens Theorem 6.3.3 for $f \geq 2$:

Theorem 6.3.5. *Consider an m^n factorial. Let $1 \leq u \leq m^f + 1$ and $d_0(\in \mathcal{D}_{N_0+u})$ be obtained by adding any u treatment combinations to a plan given by a symmetric $OA(N_0, n, m, 2f)$. Then d_0 is a Resolution (f, f) plan. Furthermore d_0 is E-optimal in $\mathcal{D}_{N_0+u}$ if any of the following holds:*

(i) $m \geq 3,\ f \geq 1$,
(ii) $m = 2,\ f \geq 2$,
(iii) $m = 2,\ f = 1,\ n \geq 3$. ∎

Even for $f = 1$, it is difficult to extend Theorem 6.3.5, in full generality, to the case of asymmetric factorials. Theorem 6.3.6 below, due to Mukerjee (1995), gives a partial extension. Part (b) of this theorem shows that at least in an important special case a full extension of Theorem 6.3.5 is possible.

Theorem 6.3.6. *Consider an $m_1 \times \cdots \times m_n$ factorial where $m_1 \geq \cdots \geq m_n (\geq 2)$ and $m_1, \ldots, m_n$ are not all equal. Suppose that there exists an orthogonal array $OA(N_0, n, m_1 \times \cdots \times m_n, 2f)$.*

(a) *For* $1 \le u \le (\prod_{i=1}^{f} m_i) - 1$, *let* $d_0(\in \mathcal{D}_{N_0+u})$ *be obtained by adding any* u *treatment combinations to the plan given by the array. Then* d_0 *is a Resolution* (f, f) *plan that is* E*-optimal in* $\mathcal{D}_{N_0+u}$.

(b) *If, in addition,* $m_1 = \cdots = m_{2f} = m$, *then for* $1 \le u \le m^f + 1$, *a plan* $d_0(\in \mathcal{D}_{N_0+u})$ *obtained by adding any* u *treatment combinations to the array is a Resolution* (f, f) *plan that is* E*-optimal in* $\mathcal{D}_{N_0+u}$. ■

Example 6.3.3. In Sections 4.3 and 4.5 (see Examples 4.3.1, 4.5.2 and the discussion following Theorem 4.5.1), we have noted that the following orthogonal arrays exist: (i) $OA(8, 5, 4 \times 2^4, 2)$, (ii) $OA(18, 7, 3^6 \times 6, 2)$, (iii) $OA(16, 11, 4^2 \times 2^9, 2)$. By Theorem 6.3.6, plans obtained by adding up to any three, five, or five runs, respectively to the arrays in (i), (ii), and (iii) are E-optimal Resolution (1, 1) plans within the relevant classes. ■

Remark 6.3.3. From Theorems 6.3.5 and 6.3.6, it is clear that orthogonal array plus one run plans are always E-optimal even for asymmetric factorials. This may be contrasted with what was seen earlier in Example 6.2.1 under the A- and D-criteria. It will be of interest to examine if these theorems can be strengthened further leading to increased upper bounds for u. This, however, seems to be a difficult problem even for $f = 1$. ■

Remark 6.3.4. In the setup of Theorem 6.3.5 or 6.3.6, optimal designs under other commonly used criteria, like the D- or A-criteria, are often as yet unknown. As for the performance of an E-optimal design d_0 considered in either of these theorems under such criteria, it is intuitively clear that d_0 should be highly efficient if u is small compared to N_0. Even otherwise, numerical studies in Mukerjee (1995), based on comparison of $\det({\mathcal{I}_{d_0}}^{-1})$ or $\mathrm{tr}({\mathcal{I}_{d_0}}^{-1})$ with certain conservative (possibly unattainable) lower bounds, suggest that if the columns of $P^{(1)}$ corresponding to the u added runs are mutually orthogonal or at least linearly independent, then d_0 tends to behave quite satisfactorily under the D- or A-criteria. ■

Before concluding this section, we present another result that itself is not of interest but will be useful in the next section. Let ${P_*}^{(1)}$ be an $(\alpha_f - 1) \times v$ submatrix of $P^{(1)}$ obtained by deleting the row $P^{00\ldots0}$ of the latter, e.g., if $f = 1$ then (see (6.3.1))

$$ {P_*}^{(1)} = (P^{10\ldots0'}, P^{01\ldots0'}, \ldots, P^{00\ldots1'})'. $$

For any $d \in \mathcal{D}_N$, let

$$ {\mathcal{I}_d}^* = {P_*}^{(1)} R_d ({P_*}^{(1)})'. \tag{6.3.36} $$

A plan $d_0 (\in \mathcal{D}_N)$ will be said to be E^*-optimal in $\mathcal{D}_N$ if $\mathcal{I}_{d_0}{}^*$ is positive definite and d_0 maximizes $\mu_{\min}(\mathcal{I}_d{}^*)$ over $\mathcal{D}_N$. As in Lemma 2.5.1, for each $d \in \mathcal{D}_N$,

$$\mathrm{tr}(\mathcal{I}_d{}^*) = \frac{N}{v}(\alpha_f - 1).$$

Hence from (2.6.11) it is not hard to see that if $d_0 (\in \mathcal{D}_N)$ is represented by an $OA(N, n, m_1 \times \cdots \times m_n, 2f)$, then d_0 is E^*-optimal in $\mathcal{D}_N$. Furthermore, analogously to Theorem 6.3.5, the following result, reported in Mukerjee (1995), holds:

Theorem 6.3.7. *Consider an m^n factorial. Let $1 \le u \le m^f$ and $d_0 (\in \mathcal{D}_{N_0+u})$ be obtained by adding any u treatment combinations to a plan given by a symmetric $OA(N_0, n, m, 2f)$. Then d_0 is E^*-optimal in $\mathcal{D}_{N_0+u}$ if any one of the following holds:*

(i) $m \ge 3,\ f \ge 1$.
(ii) $m = 2,\ f \ge 2$.
(iii) $m = 2,\ f = 1, n \ge 3$. ■

6.4 NEARLY ORTHOGONAL ARRAYS

With reference to situations where the nature of $N, n, m_1, \ldots, m_n$ rules out the existence of an $OA(N, n, m_1 \times \cdots \times m_n, 2)$, Wang and Wu (1992) proposed a modification of the method of construction described in Section 4.4 to get what they called nearly orthogonal arrays. The resulting fractional factorial plans are almost saturated and not obtainable by directly augmenting orthogonal arrays. Later in this section we present an outline of the construction procedure of Wang and Wu (1992). From there, one can note that their nearly orthogonal arrays are often quite close to orthogonal arrays. Hence the associated fractional factorial plans are intuitively expected to behave well under standard optimality criteria. This section aims at showing that such plans are indeed, E-optimal in many situations. Theorem 6.4.1 below will play a key role for this purpose. Throughout this section, $f = 1$; namely, the model involves only the general mean and the main effects. We will find it convenient in this section to use the more explicit notation $\mathcal{D}_N(m_1 \times \cdots \times m_n)$ for the class of all N-run plans for an $m_1 \times \cdots \times m_n$ factorial. As in Chapter 1 the n factors will be denoted by $F_1, \ldots, F_n$.

Consider any plan $d \in \mathcal{D}_N(m_1 \times \cdots \times m_n)$. In analogy with (6.3.23) and (6.3.24), for $1 \le i \le n$, let R_{id} be an $m_i \times m_i$ diagonal matrix with diagonal entries given by the frequencies with which the levels of F_i occur in d, and $\boldsymbol{r}_{id} = R_{id}\mathbf{1}_{m_i}$. Also, for $1 \le i, u \le n, i \ne u$, let $B_d{}^{(i,u)}$ be an $m_i \times m_u$ matrix with (j_i, j_u)th entry representing the number of times the level j_i of F_i occurs with the level j_u of F_u in d $(0 \le j_i \le m_i - 1, 0 \le j_u \le m_u - 1)$. Then by (6.1.1), just as in (6.3.22), one gets

$$\mathcal{I}_d = v^{-1} P_0 \begin{bmatrix} N & \boldsymbol{r}_{1d}{}' & \boldsymbol{r}_{2d}{}' & \dots & \boldsymbol{r}_{nd}{}' \\ \boldsymbol{r}_{1d} & R_{1d} & B_d{}^{(1,2)} & \dots & B_d{}^{(1,n)} \\ \boldsymbol{r}_{2d} & B_d{}^{(2,1)} & R_{2d} & \dots & B_d{}^{(2,n)} \\ \vdots & & & & \\ \boldsymbol{r}_{nd} & B_d{}^{(n,1)} & B_d{}^{(n,2)} & \dots & R_{nd} \end{bmatrix} P_0{}', \qquad (6.4.1)$$

where $v = \prod_{i=1}^{n} m_i$ and

$$P_0 = \mathrm{diag}(1, m_1{}^{1/2} P_1, \dots, m_n{}^{1/2} P_n). \qquad (6.4.2)$$

Now let $s(< n)$ be a positive integer and $N = N_1 N_2$ where $N_1, N_2(> 1)$ are integers. Let $d_{01} \in \mathcal{D}_N(m_1 \times \cdots \times m_s)$ be such that it can be partitioned into N_1 mutually exclusive and exhaustive parts, each consisting of N_2 runs such that for $1 \le i \le s$, the levels of F_i appear equally often in each part. Then N_2 is an integral multiple of each of $m_1, \dots, m_s$. For $1 \le h \le N_1$, let the treatment combinations in the hth part of d_{01} be $\psi_{h1}, \dots, \psi_{hN_2}$. Let $\hat{d_{02}} \in \mathcal{D}_{N_1}(m_{s+1} \times \cdots \times m_n)$ consist of treatment combinations $\hat{\psi}_1, \dots, \hat{\psi}_{N_1}$. Finally, let $d_0 \in \mathcal{D}_N(m_1 \times \cdots \times m_n)$ be given by

$$d_0 = \{\psi_{hh'}\hat{\psi}_h : 1 \le h \le N_1,\ 1 \le h' \le N_2\}. \qquad (6.4.3)$$

From each treatment combination in d_0, if the levels of $F_{s+1}, \dots, F_n$ are deleted, then one gets the plan d_{01}; on the other hand, if the levels of $F_1, \dots, F_s$ are deleted, then one gets the plan $d_{02} \in \mathcal{D}_N(m_{s+1} \times \cdots \times m_n)$ which is an N_2-fold repetition of $\hat{d_{02}}$. Then the following result, due to Mukerjee (1995), holds:

Theorem 6.4.1. *With reference to the setup described in the last paragraph, suppose that d_{01} is E^*-optimal in $\mathcal{D}_N(m_1 \times \cdots \times m_s)$ and that d_{02} is a Resolution $(1, 1)$ plan that is E-optimal in $\mathcal{D}_N(m_{s+1} \times \cdots \times m_n)$. Then d_0 is a Resolution $(1, 1)$ plan that is E-optimal in $\mathcal{D}_N(m_1 \times \cdots \times m_n)$.*

Proof. For any $d \in \mathcal{D}_N(m_1 \times \cdots \times m_n)$, one can define $d_1(\in \mathcal{D}_N(m_1 \times \cdots \times m_s))$ obtained by deleting the levels of $F_{s+1}, \dots, F_n$ from each treatment combination in d, and $d_2(\in \mathcal{D}_N(m_{s+1} \times \cdots \times m_n))$ obtained by deleting the levels of $F_1, \ldots, F_s$ from each treatment combination in d. From (6.3.36) and (6.1.1), analogously to (6.4.1),

$$\mathcal{I}_{d_1}{}^* = v_1{}^{-1} P_{01} \begin{bmatrix} R_{1d} & B_d{}^{(1,2)} & \ldots & B_d{}^{(1,s)} \\ B_d{}^{(2,1)} & R_{2d} & \ldots & B_d{}^{(2,s)} \\ \vdots & & & \\ B_d{}^{(s,1)} & B_d{}^{(s,2)} & \ldots & R_{sd} \end{bmatrix} P_{01}{}', \quad (6.4.4)$$

$$\mathcal{I}_{d_2} = v_2{}^{-1} P_{02} \begin{bmatrix} N & \boldsymbol{r}_{s+1\,d}{}' & \ldots & \boldsymbol{r}_{nd}{}' \\ \boldsymbol{r}_{s+1\,d} & R_{s+1\,d} & \ldots & B_d{}^{(s+1,n)} \\ \vdots & & & \\ \boldsymbol{r}_{nd} & B_d{}^{(n,s+1)} & \ldots & R_{nd} \end{bmatrix} P_{02}{}', \quad (6.4.5)$$

where $v_1 = \prod_{i=1}^{s} m_i$, $v_2 = \prod_{i=s+1}^{n} m_i$, and

$$P_{01} = \text{diag}(m_1{}^{1/2} P_1, \ldots, m_s{}^{1/2} P_s), \quad (6.4.6)$$

$$P_{02} = \text{diag}(1, m_{s+1}{}^{1/2} P_{s+1}, \ldots, m_n{}^{1/2} P_n). \quad (6.4.7)$$

By (6.4.1), (6.4.2), and (6.4.4)–(6.4.7), there exists a permutation matrix Z such that for every $d \in \mathcal{D}_N(m_1 \times \cdots \times m_n)$,

$$Z\mathcal{I}_d Z' = v^{-1} \begin{bmatrix} v_2 \mathcal{I}_{d_2} & C_d \\ C_d{}' & v_1 \mathcal{I}_{d_1}{}^* \end{bmatrix}, \quad (6.4.8)$$

where

$$C_d = P_{02} \begin{bmatrix} \boldsymbol{r}_{1d}{}' & \ldots & \boldsymbol{r}_{sd}{}' \\ B_d{}^{(s+1,1)} & \ldots & B_d{}^{(s+1,s)} \\ \vdots & & \\ B_d{}^{(n,1)} & \ldots & B_d{}^{(n,s)} \end{bmatrix} P_{01}{}'. \quad (6.4.9)$$

By (6.4.8), for each $d \in \mathcal{D}_N(m_1 \times \cdots \times m_n)$,

$$\mu_{\min}(\mathcal{I}_d) \le v^{-1} \min\{v_2 \mu_{\min}(\mathcal{I}_{d_2}), v_1 \mu_{\min}(\mathcal{I}_{d_1}{}^*)\}. \quad (6.4.10)$$

By our construction, for $1 \le i \le s$, $s + 1 \le u \le n$, each row vector of $B_{d_0}{}^{(u,i)}$ is proportional to $\mathbf{1}'_{m_i}$. Furthermore, for $1 \le i \le s$, the levels of F_i occur equally often in d_{01} and hence in d_0. Therefore, by (2.2.2), (6.4.6), and

(6.4.9), C_{d_0} is a null matrix so that by (6.4.8),

$$\mu_{\min}(\mathcal{I}_{d_0}) = v^{-1}\min\{v_2\mu_{\min}(\mathcal{I}_{d_{02}}), v_1\mu_{\min}(\mathcal{I}_{d_{01}}{}^*)\}.$$

The result now follows from (6.4.10) and the given conditions on d_{01} and d_{02}. ■

We now discuss how a substantial part of the constructions due to Wang and Wu (1992) can be linked with (6.4.3). The notion of a difference matrix was introduced in Definition 3.5.1. In the same spirit, a *nearly difference matrix* is a matrix with elements from a finite, additive Abelian group $\mathcal{G}$ such that among the entrywise differences of elements of every two distinct columns, the elements of $\mathcal{G}$ occur as evenly as possible. For example, given a difference matrix, the augmentation of any row with elements from the same group yields a nearly difference matrix. Now, with reference to (6.4.3), let $s \geq 2$, $m_1 = \cdots = m_s = m(\geq 2)$, and suppose that d_{01} consists of treatment combinations given by the rows of $\boldsymbol{\gamma} \odot \Delta$, where Δ is an $N_1 \times s$ difference matrix or nearly difference matrix with elements from an additive group $\mathcal{G} = \{\gamma_0, \gamma_1, \ldots, \gamma_{m-1}\}$, $\boldsymbol{\gamma} = (\gamma_0, \gamma_1, \ldots, \gamma_{m-1})'$, and as before, $\odot$ stands for Kronecker sum. Then $d_{01} \in \mathcal{D}_N(m^s)$, where $N = N_1 m$ and d_{01} can be partitioned into N_1 parts, each consisting of m runs, such that in each part the levels of every factor involved in d_{01} appear equally often. Note that each row of Δ accounts for one of these parts. Next suppose that the treatment combinations in $\hat{d}_{02} \in \mathcal{D}_{N_1}(m_{s+1} \times \cdots \times m_n)$ are represented by the rows of an $N_1 \times (n - s)$ array L. Finally, following Wang and Wu (1992), let $d_0 \in \mathcal{D}_N(m^s \times m_{s+1} \times \cdots \times m_n)$ consist of treatment combinations given by the rows of

$$[\boldsymbol{\gamma} \odot \Delta \,\vdots\, \mathbf{0}_m \odot L], \tag{6.4.11}$$

where $\mathbf{0}_m$ is an $m \times 1$ null vector. Note that (6.4.11) is in conformity with (6.4.3) and that, with reference to (6.4.11), d_{02} is given by the $N \times (n - s)$ array $\mathbf{0}_m \odot L$. Since Δ is a difference or a nearly difference matrix, a comparison of (6.4.11) with the construction procedure described in Section 4.4 suggests that if L is close to an orthogonal array, then so will be the structure given by (6.4.11). In this sense, Wang and Wu (1992) referred to (6.4.11) as a nearly orthogonal array.

Theorem 6.4.1 is applicable in proving the E-optimaity of d_0, given by (6.4.11), provided that d_{01}, given by $\boldsymbol{\gamma} \odot \Delta$, is E^*-optimal and d_{02}, given by $\mathbf{0}_m \odot L$, is E-optimal in relevant classes. If Δ is a difference matrix, then d_{01} is given by an orthogonal array of strength two, and therefore is E^*-optimal. On the other hand, if Δ is a nearly difference matrix obtained by augmenting any single row to an $(N_1 - 1) \times s$ difference matrix, then d_{01} consists of an

m-symbol orthogonal array of strength two together with m extra runs and hence, by Theorem 6.3.7, is E^*-optimal if either $m \geq 3$, $s \geq 2$ or $m = 2$, $s \geq 3$; see parts (a) and (d) of Example 6.4.1 below for an illustration. Again, if $\hat{d}_{02}$ is given by an orthogonal array of strength two, then so is d_{02}, and hence d_{02} is E-optimal. On the other hand, if the array L representing $\hat{d}_{02}$ is obtained by the augmentation of $\hat{u}$ runs to an orthogonal array of strength two, then d_{02} consists of an orthogonal array of strength two together with $\hat{u}m$ extra runs, and if $\hat{u}m$ is sufficiently small, then Theorem 6.3.5 or 6.3.6 can be employed to prove the E-optimality of d_{02}; see parts (b), (c), and (e) of Example 6.4.1 below for illustration.

Following Mukerjee (1995), some representative cases are illustrated in Example 6.4.1 below. In all these cases, d_0 is constructed via (6.4.11). We only present Δ and L and indicate the results needed in addition to Theorem 6.4.1 to show the E-optimality of d_0. All the difference matrices mentioned in Example 6.4.1, except the one considered in part (b), arise from Hadamard matrices as discussed in Section 3.5.

Example 6.4.1. The following yield Resolution (1, 1) plans that are E-optimal in relevant classes.

- **(a)** (18-run plan for $2^8 \times 3^4$ factorial): Take Δ as a 9×8 nearly difference matrix obtained by the augmentation of any single row to a difference matrix $D(8, 8; 2)$. Let L be a symmetric $OA(9, 4, 3, 2)$ as shown in Example 2.6.1. Use Theorem 6.3.7.
- **(b)** (27-run plan for $3^9 \times 4 \times 2^4$ factorial): Take $\Delta = D(9, 9; 3)$ as given by Lemma 3.5.1 and obtain L by adding any single run to an $OA(8, 5, 4 \times 2^4, 2)$ as shown in Example 2.6.1. Use Theorem 6.3.6.
- **(c)** (40-run plan for $2^{20} \times 3^6 \times 6$ factorial): Take $\Delta = D(20, 20; 2)$ and obtain L by adding any two runs to an $OA(18, 7, 3^6 \times 6, 2)$ as shown in Example 4.5.2. Use Theorem 6.3.6.
- **(d)** (50-run plan for $2^{24} \times 5^6$ factorial): Take Δ as a 25×24 nearly difference matrix obtained by the augmentation of any single row to $D(24, 24; 2)$. Let L be a symmetric $OA(25, 6, 5, 2)$ arising from Theorem 3.4.1(a). Use Theorem 6.3.7.
- **(e)** (56-run plan for $2^{28} \times 3^{13}$ factorial): Take $\Delta = D(28, 28; 2)$ and obtain L by adding any single run to a symmetric $OA(27, 13, 3, 2)$ arising from Theorem 3.4.3. Use Theorem 6.3.5. ■

No result is as yet available on the D- or A-optimality of plans based on nearly orthogonal arrays. However, the numerical studies reported in Mukerjee (1995) indicate that such plans can be highly efficient under these criteria as well.

6.5 CONNECTION WITH WEIGHING DESIGNS

In Section 6.3 we had hinted at a connection between optimal chemical balance weighing designs and optimal fractions of 2^n factorials. We now examine this point in some more detail and indicate a few related developments. In view of (6.2.4), the chemical balance weighing designs that are relevant in the present context are those for which the elements of the design matrix are ± 1. Let $\mathcal{X}(N, u)$ be the class of $N \times u$ matrices X with elements ± 1 and $\mathcal{X}^*(N, u)$ be a subclass thereof consisting of those matrices X for which $X'X$ is positive definite. An $X_0 (\in \mathcal{X}(N, u))$, representing a chemical balance weighing design, is said to be D-, A-, or E-optimal in $\mathcal{X}(N, u)$ if X_0 belongs to $\mathcal{X}^*(N, u)$ and minimizes $\det(X'X)^{-1}$ or $\text{tr}(X'X)^{-1}$ or $\mu_{\max}\{(X'X)^{-1}\}$, respectively, over $X \in \mathcal{X}^*(N, u)$. Since $X_0{}'X_0$ remains unaltered when any row of X_0 is multiplied by -1, without loss of generality, it may be supposed that the first column of X_0 consists of only $+1$'s.

Let J_{Nu} be the $N \times u$ matrix with elements unity, and $n = u - 1$. We revert back to the notation $\mathcal{D}_N$ in this section to denote the class of N-run plans.

Theorem 6.5.1. *Let $X_0 \in \mathcal{X}(N, u)$ be D- (or A- or E-) optimal in $\mathcal{X}(N, u)$, and without loss of generality, suppose that the first column of X_0 consists of only $+1$'s. Let*

$$\bar{X}_0 = \frac{1}{2}(X_0 + J_{Nu}), \tag{6.5.1}$$

and d_0 be an N-run fraction of a 2^n factorial consisting of treatment combinations given by the rows of the $N \times n$ array obtained by deleting the first column of $\bar{X}_0$. Then d_0 is a Resolution (1, 1) plan that is D- (or A- or E-, respectively) optimal in $\mathcal{D}_N$.

Proof. First note that the elements of $\bar{X}_0$ are 0 or 1, so d_0 is well defined as a member of $\mathcal{D}_N$. For any $d \in \mathcal{D}_N$, define the matrices G_d and K_d as in the proof of Theorem 6.3.1. Then, with $v = 2^n$, by (6.3.4) and (6.3.5),

$$\mathcal{I}_d = v^{-1} K_d K_d{}' \tag{6.5.2}$$

where $K_d{}' \in \mathcal{X}(N, u)$, as $u = n + 1$. With $P^{(1)}$ as in (6.3.1) and $P_i = (-\frac{1}{\sqrt{2}}, \frac{1}{\sqrt{2}})$ for each i (see Remark 2.3.1), it is clear that if $j_1 \ldots j_n$ is any typical treatment combination in d, then the corresponding row of $K_d{}'$ is $(1, 2j_1 - 1, \ldots, 2j_n - 1)$. Now by (6.5.1), $X_0 = 2\bar{X}_0 - J_{Nu}$. Hence recalling that the first column of X_0 consists of only $+1$'s, from the definition of d_0 we get $K_{d_0} = X_0'$; namely,

$$\mathcal{I}_{d_0} = v^{-1} X_0' X_0. \qquad (6.5.3)$$

In view of the stated optimality of X_0, the result now follows from (6.5.2) and (6.5.3). ■

Because of Theorem 6.5.1, every optimality result over $\mathcal{X}(N, u)$ has a counterpart in terms of a 2^n factorial. An attempt to review all such optimality results, several of which are specific to particular N or u, will lead to considerable digression. We therefore refrain from doing so and present only two illustrative examples, the first of which is quite general in nature. The interested reader may refer to Ehlich (1964), Yang (1968), Galil and Kiefer (1980, 1982), Kounias and Chadjipantelis (1983), Cheng et al. (1985), Chadjipantelis and Kounias (1985), Sathe and Shenoy (1989, 1990, 1991), Farmakis (1992), Moyssiadis et al. (1995), Koukouvinos (1996), and the references therein for further details. Some of these authors (e.g., Sathe and Shenoy, 1990, 1991) provide extensive tables of optimal designs over $\mathcal{X}(N, u)$. Chapter 8 of Shah and Sinha (1989) contains an informative summary of the developments in this area up to that stage.

Example 6.5.1. Let $N \equiv 3 \bmod 4$, $u \leq N$ and $n = u - 1$. Suppose that there exists $X_0 \in \mathcal{X}(N, u)$ such that

$$X_0' X_0 = (N + 1) I_u - J_{uu}. \qquad (6.5.4)$$

Clearly, $X_0' X_0$ is positive definite. Furthermore, following Galil and Kiefer (1980), X_0 is D-optimal in $\mathcal{X}(N, u)$ provided that

$$N \geq 2u - 5. \qquad (6.5.5)$$

Also, as shown by Sathe and Shenoy (1989), X_0 is A-optimal in $\mathcal{X}(N, u)$ provided that

$$N \geq \tfrac{1}{4}[7u - 16 + \{(u - 4)(17u - 36)\}^{1/2}] \quad \text{and} \quad u \geq 4. \qquad (6.5.6)$$

Thus a Resolution (1, 1) plan d_0 constructed from X_0 following Theorem 6.5.1 will be D- or A-optimal in $\mathcal{D}_N$, as a fraction of a 2^n factorial, under (6.5.5) or (6.5.6), respectively. ■

Remark 6.5.1.

(a) A matrix X_0, as envisaged in (6.5.4), is always available if a Hadamard matrix H_{N+1} of order $N+1$ exists. Without loss of generality, let the first row and first column of H_{N+1} consist only of $+1$'s. Then X_0, satisfying (6.5.4), can be obtained by deleting the first row and any $N+1-u$ of the last N columns of H_{N+1}.

(b) Under the above construction, the first column of X_0 contains only $+1$'s, and in consideration of Theorem 6.5.1 and the discussion foregoing Theorem 3.2.2, the addition of the run $11\ldots1$ to the plan d_0 arising from X_0 will produce a symmetric $OA(N+1, n, 2, 2)$. The reader may therefore wonder if an analogue of Theorem 6.2.1 holds for orthogonal array minus one-run plans at least when $f = 1$ and both (6.5.5) and (6.5.6) hold. The answer to this question is, however, in the negative. To that effect, let

$$4 < u(= n+1) \leq N-3, \tag{6.5.7}$$

and suppose that a Hadamard matrix H_{N-3} also exists. Then by Theorem 3.2.2, one can get an $OA(N-3, N-4, 2, 2)$ and hence an $OA(N-3, n, 2, 2)$. Let $d_1 (\in \mathcal{D}_N)$ be obtained by the addition of any three runs to the latter array. By Theorem 6.3.3, d_1 is E-optimal in $\mathcal{D}_N$. To see that d_0 is not so, note that by (6.3.13) and (6.3.14), $\mu_{\min}(\mathcal{I}_{d_1}) = (N-3)/v$, while by (6.5.3) and (6.5.4), $\mu_{\min}(\mathcal{I}_{d_0}) = (N+1-u)/v$. Use of (6.5.7) now yields

$$\mu_{\min}(\mathcal{I}_{d_1}) > \mu_{\min}(\mathcal{I}_{d_0}), \tag{6.5.8}$$

showing that d_1 dominates d_0 under the E-criterion even when (6.5.5) and (6.5.6) hold. We emphasize that there are plenty of combinations (N, u), like (11,5), (11,6), (15,6), (15,7), and so on, for which (6.5.5)–(6.5.7) hold and both H_{N+1} and H_{N-3} exist. Because of (6.5.8), for no such (N, u), the orthogonal array minus one-run plan d_0 can be optimal in $\mathcal{D}_N$ in the sense of Theorem 6.2.1 notwithstanding its optimality under the D- and A-criteria. More discussion on the issue of addition vis-à-vis deletion of runs with reference to a two-symbol orthogonal array is available in Dey and Midha (1998). ■

Example 6.5.2. Let $N = u = 7$ and $n = 6$. Then neither (6.5.5) nor (6.5.6) holds, and following Galil and Kiefer (1980), Sathe and Shenoy (1989), and Bagiatis (1990),

$$X_0 = \begin{bmatrix} 1 & -1 & -1 & -1 & -1 & -1 & -1 \\ 1 & 1 & -1 & 1 & 1 & -1 & -1 \\ 1 & -1 & 1 & 1 & 1 & -1 & -1 \\ 1 & -1 & -1 & 1 & -1 & 1 & 1 \\ 1 & -1 & -1 & -1 & 1 & 1 & 1 \\ 1 & 1 & 1 & -1 & -1 & -1 & 1 \\ 1 & 1 & 1 & -1 & -1 & 1 & -1 \end{bmatrix}$$

is D-, A- and E-optimal in $\mathcal{X}(7, 7)$. Following Theorem 6.5.1, X_0 yields the fraction

$$d_0 = \{000000, 101100, 011100, 001011, 000111, 110001, 110010\}$$

of a 2^6 factorial. Clearly, d_0 represents a Resolution (1, 1) plan that is D-, A-, and E-optimal in $\mathcal{D}_7$. Here $n = N - 1$, and therefore it is easy to verify that an E-optimal plan cannot be obtained via Theorem 6.3.3 (see Remark 6.5.2 below).

In the spirit of Remark 6.5.1(b), it is of interest to compare d_0 with a rival plan $d_1 (\in \mathcal{D}_7)$ obtained by the deletion of any one run from an $OA(8, 6, 2, 2)$. Upon explicit computation of $X_0{}'X_0$ and use of (6.5.3), it can be seen that the eigenvalues of $\mathcal{I}_{d_0}$ are $4/v$, $9/v$, and $12/v$, with respective multiplicities 4, 1 and 2, where $v = 2^6$. Similarly, from an analogue of (6.2.2), one can check that the eigenvalues of $\mathcal{I}_{d_1}$ are $1/v$ and $8/v$ with respective multiplicities 1 and 6. Hence

$$\det(\mathcal{I}_{d_0}) = \frac{331776}{v^7}, \qquad \det(\mathcal{I}_{d_1}) = \frac{262144}{v^7},$$
$$\operatorname{tr}(\mathcal{I}_{d_0}{}^{-1}) = 1.278v, \qquad \operatorname{tr}(\mathcal{I}_{d_1}{}^{-1}) = 1.75v,$$
$$\mu_{\min}(\mathcal{I}_{d_0}) = \frac{4}{v}, \qquad \mu_{\min}(\mathcal{I}_{d_1}) = \frac{1}{v}.$$

This shows the superiority of d_0 over d_1 under each of D-, A-, and E-criteria and reinforces what was stated in Remark 6.5.1(b). ■

Remark 6.5.2. At this stage it is appropriate to summarize the main tools that are helpful in identifying optimal Resolution (1, 1) plans for 2^n factorials in $\mathcal{D}_N$. Note that $f = 1$ and that, as mentioned in Section 6.1, $N \geq \alpha_f = n + 1$, namely, $n \leq N - 1$. To avoid trivialities, let $n \geq 3$.

(a) If $N \equiv 0 \bmod 4$ and a Hadamard matrix of order N, (i.e., H_N exists), then, by Theorem 3.2.2, an $OA(N, n, 2, 2)$ is available that, following Theorem 2.6.1, gives a universally optimal plan.

(b) Let $N \equiv 1 \bmod 4$. If a Hadamard matrix H_{N-1} exists and $n \leq N - 2$, then as in (a), an $OA(N - 1, n, 2, 2)$ is available and use of Theorem 6.2.1 yields a plan that is optimal with respect to every generalized criterion of type 1.

(c) Let $N \equiv 2 \bmod 4$. If a Hadamard matrix H_{N-2} exists and $n \leq N - 3$, then an $OA(N - 2, n, 2, 2)$ is available, and from Theorem 6.3.1 one gets a plan that is optimal with respect to every type 1 criterion satisfying $\lim_{\mu \to 0^+} q(\mu) = q(0) = +\infty$ and also the E-criterion (*vide* Remark 6.3.2).

(d) Let $N \equiv 3 \bmod 4$. If a Hadamard matrix H_{N-3} exists and $n \leq N - 4$, then an $OA(N - 3, n, 2, 2)$ is available and use of Theorem 6.3.3 yields an E-optimal plan. On the other hand, if H_{N+1} exists and (6.5.5) or (6.5.6) holds with $u = n + 1$, then a D- or A-optimal plan can be obtained via Example 6.5.1 and Remark 6.5.1(a).

Following Section 3.2, the relevant Hadamard matrices exist for practically useful values of N and (a)–(d) above, indeed, cover a large number of cases. However, there can be situations, as in Example 6.5.2, that are not covered by any of the above. For such (N, n), the references cited before Example 6.5.1 and the tables reported in such references might be of help. ■

Optimal Resolution (1, 1) plans for 2^n factorials have been studied in the literature from several other viewpoints. Raktoe and Federer (1970) and Raktoe (1974) considered characterizations for the optimal design problem in the saturated case. In particular, the latter author showed that the problem is equivalent to that of choosing a simplex in the Euclidean space. Further results of related interest were reported in Raktoe and Federer (1973). Kounias et al. (1983) and Kolyva-Machera (1989b) worked under the G-criterion where the objective is to choose a design so as to minimize the maximum variance of the estimated response across all treatment combinations. Bagiatis (1990) proposed an algorithm for searching a plan that is at least nearly E-optimal. Masaro and Wong (1992) considered certain criteria that are of type 1 but different from the usual D- or A-criteria.

The design problem for 2^n factorials, when the model includes the general mean, all main effects, and one or more specified two-factor interaction(s), was considered by several authors. To give a flavor of such work, we present a result following Hedayat and Pesotan (1992). Let $N = 8s$, $n = 8s - 2$, and suppose that the model includes one specific two-factor interaction, say without loss of generality, that between the first and $(4s)$th factors, in addition to the general mean and the main effects. If interest lies in all these, then, as in (6.1.1), the information matrix under any $d(\in \mathcal{D}_N)$ is given by

$$\tilde{\mathcal{I}}_d = \tilde{P} R_d \tilde{P}', \tag{6.5.9}$$

where

$$\tilde{P} = (P^{00\ldots0'}, P^{10\ldots0'}, P^{01\ldots0'}, \ldots, P^{00\ldots1'}, P^{z'})', \tag{6.5.10}$$

and z is a binary n-tuple with 1 in the first and $(4s)$th positions and zero elsewhere. Since $\tilde{P}$ has $n + 2(= N)$ rows, analogously to (6.1.2), for every $d \in \mathcal{D}_N$,

$$\text{tr}(\tilde{\mathcal{I}}_d) = \frac{N^2}{v}, \tag{6.5.11}$$

where $v = 2^n$. Now suppose that a Hadamard matrix

$$H = [\mathbf{1}_{4s} \vdots M], \tag{6.5.12}$$

of order $4s$ is available. Let

$$M_0 = \begin{bmatrix} M & M \\ M & -M \end{bmatrix}, \tag{6.5.13}$$

and consider a plan $d_0(\in \mathcal{D}_N)$ such that the treatment combinations included in d_0 are given by the rows of M_0 with the -1's replaced by zeros (see (6.5.1)). As in the proof of Theorem 6.5.1, from (2.2.4), (6.5.9), (6.5.10), and (6.5.13), one can check that

$$\tilde{\mathcal{I}}_{d_0} = v^{-1} K_{d_0} K_{d_0}', \tag{6.5.14}$$

where

$$K_{d_0} = \begin{bmatrix} \mathbf{1}_{4s} & M & M & \mathbf{1}_{4s} \\ \mathbf{1}_{4s} & M & -M & -\mathbf{1}_{4s} \end{bmatrix}'. \tag{6.5.15}$$

By (6.5.12), (6.5.14), and (6.5.15), $\tilde{\mathcal{I}}_{d_0} = (N/v)I_N$. Hence, in view of (6.5.11), analogously to Lemma 2.5.2 the following result holds:

Theorem 6.5.2. *Under the setup described above, the plan d_0 is universally optimal in $\mathcal{D}_N$ for the estimation of the general mean, all main effects and the specified two-factor interaction.* ■

It is easy to see that with one or more column(s) of M_0, other than the first or $(4s)$th columns, deleted, the resulting plan involves a fewer number of factors and enjoys the same optimality property. For further developments on the design problem with specific two-factor interactions in the model, the reader is referred to Wu and Chen (1992), Hedayat and Pesotan (1992, 1997), Vijayan and Shah (1997), and the references therein.

6.6 OPTIMALITY WITH TWO OR THREE FACTORS

There are certain special features inherent in the study of optimal saturated or nearly saturated Resolution (1, 1) plans when the number of factors, n, equals two or three. For example, as illustrated in Theorem 6.6.1 below, unimodular matrices, well-known in the context of the transportation problem in operations research, can play a key role. In this section we look at some of these aspects. Throughout this section, $f = 1$; namely, the model involves only the general mean and the main effects.

First suppose that $n = 2$, and consider the saturated case given by $N = \alpha_f = m_1 + m_2 - 1$ (see Theorem 2.3.2). Then the following result, due to Mukerjee et al. (1986) and Krafft (1990), holds. The result was also obtained by Pesotan and Raktoe (1988) in the special case $m_1 = m_2$ as a part of a wider study.

Theorem 6.6.1. *Let $n = 2$ and $N = m_1 + m_2 - 1$. Then all saturated Resolution* (1, 1) *plans are equivalent under the D-criterion, and hence all of them are D-optimal in $\mathcal{D}_N$.*

Proof. With $n = 2$, as in (6.3.20),

$$P^{(1)} = v^{-1/2} P_0 Q_0 \tag{6.6.1}$$

where $v = m_1 m_2$, $P_0 = \text{diag}(1, {m_1}^{1/2} P_1, {m_2}^{1/2} P_2)$, and

$$Q_0 = \begin{bmatrix} \mathbf{1}'_{m_1} \otimes \mathbf{1}'_{m_2} \\ I_{m_1} \otimes \mathbf{1}'_{m_2} \\ \mathbf{1}'_{m_1} \otimes I_{m_2} \end{bmatrix}. \tag{6.6.2}$$

Let Q be an $(m_1 + m_2 - 1) \times v$ submatrix of Q_0 given by the last $m_1 + m_2 - 1$ rows of the latter. Then Q has full row rank. Moreover the rows of Q span those of Q_0 and hence, by (6.6.1), those of $P^{(1)}$. But by Lemma 2.2.1 and (2.3.9), $P^{(1)}$ has full row rank. Therefore

$$P^{(1)} = Q_1 Q \tag{6.6.3}$$

for some nonsingular matrix Q_1 of order $m_1 + m_2 - 1$.

The v columns of Q, like those of $P^{(1)}$, correspond to the lexicographically ordered treatment combinations. For any $d \in \mathcal{D}_N$, with $N = m_1 + m_2 - 1$, let Q_d be a square matrix of order $m_1 + m_2 - 1$ such that for $1 \leq i \leq m_1 + m_2 - 1$, if the ith run in d is given by the treatment combination $j_1 j_2$, then the ith column of Q_d equals the column of Q corresponding to $j_1 j_2$. By (6.1.1) and (6.6.3), then $\mathcal{I}_d = Q_1 Q_d Q_d' Q_1'$, so that

$$\det(\mathcal{I}_d) = \{\det(Q_1)\det(Q_d)\}^2. \tag{6.6.4}$$

By (6.6.2) and the definition of Q, following Hadley (1962, pp. 276–277), the matrix Q is unimodular; that is, all its minors are either 0 or ± 1. Hence, if the columns of Q_d are distinct, then $\det(Q_d) = 0$ or ± 1. On the other hand, if the columns of Q_d are not distinct, then $\det(Q_d) = 0$. Consequently, by (6.6.4), whenever $\mathcal{I}_d$ is positive definite, $\det(\mathcal{I}_d) = \{\det(Q_1)\}^2$, and the result follows. ■

While Theorem 6.6.1 relates to the estimation of the full parametric vector $\boldsymbol{\beta}^{(1)} = (\beta_{00}, \boldsymbol{\beta}_{10}', \boldsymbol{\beta}_{01}')'$, several optimality results have been reported in the two-factor case for the situation where interest lies only in complete sets of orthonormal main effect contrasts given by $\boldsymbol{\beta}_s^{(1)} = (\boldsymbol{\beta}_{10}', \boldsymbol{\beta}_{01}')'$ (see (2.3.7)); the subscript s signifies that a subvector of $\boldsymbol{\beta}^{(1)}$ is being considered. Identification of the two main effects with the block and varietal effects in a varietal block design (e.g., see Dey, 1986, ch. 2) facilitates the derivation of some of these results. To formalize the ideas, let

$$P_s^{(1)} = \begin{bmatrix} P^{10} \\ P^{01} \end{bmatrix}, \tag{6.6.5}$$

and for any $d \in \mathcal{D}_N$, analogously to (2.3.17), define

$$\mathcal{I}_{sd} = P_s^{(1)} R_d (P_s^{(1)})' - P_s^{(1)} R_d (P^{00})' \{P^{00} R_d (P^{00})'\}^{-1} P^{00} R_d (P_s^{(1)})'. \tag{6.6.6}$$

By (2.2.4) and (2.2.5),

$$P^{00} R_d (P^{00})' = \frac{N}{v} (> 0), \tag{6.6.7}$$

which shows that $\mathcal{I}_{sd}$ is well-defined. Exactly as in Theorem 2.3.1, it can be seen that under the assumed model, which corresponds to $f = 1$, the parametric vector $\boldsymbol{\beta}_s^{(1)}$ is estimable in d if and only if $\mathcal{I}_{sd}$ is positive definite (p.d.) and that under this condition the dispersion matrix of the best linear unbiased estimator of $\boldsymbol{\beta}_s^{(1)}$ arising from d is proportional to $\mathcal{I}_{sd}^{-1}$. Hence, under the plan d, the information matrix for $\boldsymbol{\beta}_s^{(1)}$ is given by $\mathcal{I}_{sd}$ just as that for $\boldsymbol{\beta}^{(1)}$ is represented by $\mathcal{I}_d$. We emphasize here that $\mathcal{I}_{sd}$ is different from the matrix $\mathcal{I}_d^*$ introduced in Section 6.3. By (6.3.36), in the present context $\mathcal{I}_d^*$

is given by the first term in the right-hand side of (6.6.6). Thus $\mathcal{I}_d^*$ in effect ignores β_{00}, whereas $\mathcal{I}_{sd}$ corresponds to the elimination of β_{00}.

Since $P^{(1)} = ((P^{00})', (P_s^{(1)})')'$ by (2.3.9), from (6.1.1), (6.6.6), and (6.6.7) it is clear that

$$\det(\mathcal{I}_d) = \left(\frac{N}{v}\right) \det(\mathcal{I}_{sd}), \tag{6.6.8}$$

so that $\mathcal{I}_{sd}$ is p.d. if and only if $\mathcal{I}_d$ is so, that is, if and only if d is a Resolution (1, 1) plan. With reference to $\boldsymbol{\beta}_s^{(1)}$, one can now define a plan $d_0(\in \mathcal{D}_N)$ to be D_s-, A_s-, or E_s-optimal if d_0 is a Resolution (1, 1) plan and minimizes $\det(\mathcal{I}_{sd}{}^{-1})$ or $\mathrm{tr}(\mathcal{I}_{sd}{}^{-1})$ or $\mu_{\max}(\mathcal{I}_{sd}{}^{-1})$, respectively, over the class of all Resolution (1, 1) plans in $\mathcal{D}_N$.

By (6.6.8), the notions of D-optimality and D_s-optimality are equivalent in the present context, and hence Theorem 6.6.1 also relates to D_s-optimality. Turning to A_s-optimality, we have the following result, due to Mukerjee et al. (1986), in the saturated case.

Theorem 6.6.2. *Let $n = 2$, $N = m_1 + m_2 - 1$, and consider a plan $d_0 \in \mathcal{D}_N$ consisting of the treatment combinations* 00, $j_1 0$, *and* $0 j_2$ $(1 \le j_1 \le m_1 - 1$, $1 \le j_2 \le m_2 - 1)$. *Then d_0 is a Resolution* (1, 1) *plan that is A_s-optimal in $\mathcal{D}_N$.*

Proof. By (2.2.4), (2.2.5), (6.6.5), and (6.6.6), just as in (6.4.1), for any $d \in \mathcal{D}_N$ one gets

$$\mathcal{I}_{sd} = \begin{bmatrix} \mathcal{I}_{sd}(1,1) & \mathcal{I}_{sd}(1,2) \\ \mathcal{I}_{sd}(2,1) & \mathcal{I}_{sd}(2,2) \end{bmatrix}, \tag{6.6.9}$$

where

$$\mathcal{I}_{sd}(i,i) = v^{-1} m_i P_i (R_{id} - N^{-1} \boldsymbol{r}_{id} \boldsymbol{r}_{id}') P_i', \quad i = 1, 2, \tag{6.6.10}$$

$$\mathcal{I}_{sd}(1,2) = \{\mathcal{I}_{sd}(2,1)\}' = v^{-1} (m_1 m_2)^{1/2} P_1 (B_d - N^{-1} \boldsymbol{r}_{1d} \boldsymbol{r}_{2d}') P_2', \tag{6.6.11}$$

$B_d = B_d{}^{(1,2)}$, and $\boldsymbol{r}_{1d}, \boldsymbol{r}_{2d}, R_{1d}, R_{2d}$ and $B_d{}^{(1,2)}$ are as defined in the context of (6.4.1). Analogously to (6.3.25) and (6.3.26),

$$\boldsymbol{r}_{1d}' \mathbf{1}_{m_1} = \boldsymbol{r}_{2d}' \mathbf{1}_{m_2} = N, \;\; B_d \mathbf{1}_{m_2} = \boldsymbol{r}_{1d}, \; B_d' \mathbf{1}_{m_1} = \boldsymbol{r}_{2d}. \tag{6.6.12}$$

Now suppose that $d(\in \mathcal{D}_N)$ is a Resolution (1, 1) plan. Then, as discussed above, $\mathcal{I}_{sd}$ is p.d.. Therefore, by (6.6.9) and (6.6.10), for $i = 1, 2$,

$$\det\left\{\begin{pmatrix} m_i^{-1/2}\mathbf{1}'_{m_i} \\ P_i \end{pmatrix} R_{id} \begin{pmatrix} m_i^{-1/2}\mathbf{1}_{m_i} & P_i' \end{pmatrix}\right\}$$

$$= \det\begin{pmatrix} \frac{N}{m_i} & m_i^{-1/2}\boldsymbol{r}_{id}'P_i' \\ m_i^{-1/2}P_i\boldsymbol{r}_{id} & P_iR_{id}P_i' \end{pmatrix}$$

$$= \left(\frac{N}{m_i}\right)\det\left\{\left(\frac{v}{m_i}\right)\mathcal{I}_{sd}(i,i)\right\} > 0,$$

which shows that the diagonal elements of R_{id} are all positive. Since by (2.2.2),

$$P_i'P_i = I_{m_i} - m_i^{-1}\mathbf{1}_{m_i}\mathbf{1}'_{m_i}, \tag{6.6.13}$$

using (6.6.10) and (6.6.12), by actual multiplication it may be verified that

$$\{\mathcal{I}_{sd}(i,i)\}^{-1} = \left(\frac{v}{m_i}\right)P_iR_{id}^{-1}P_i' \qquad (i = 1, 2).$$

Hence by (6.6.10)–(6.6.13),

$$\mathcal{I}_{sd}(1,1) - \mathcal{I}_{sd}(1,2)\{\mathcal{I}_{sd}(2,2)\}^{-1}\mathcal{I}_{sd}(2,1) = v^{-1}m_1P_1C_{1d}P_1', \tag{6.6.14}$$

$$\mathcal{I}_{sd}(2,2) - \mathcal{I}_{sd}(2,1)\{\mathcal{I}_{sd}(1,1)\}^{-1}\mathcal{I}_{sd}(1,2) = v^{-1}m_2P_2C_{2d}P_2', \tag{6.6.15}$$

where

$$C_{1d} = R_{1d} - B_dR_{2d}^{-1}B_d', \quad C_{2d} = R_{2d} - B_d'R_{1d}^{-1}B_d. \tag{6.6.16}$$

The matrices $P_iC_{id}P_i'$, $i = 1, 2$, are p.d., since so is $\mathcal{I}_{sd}$. Hence by (6.6.9), (6.6.14), and (6.6.15),

$$\operatorname{tr}(\mathcal{I}_{sd}^{-1}) = \sum_{i=1}^{2}\left(\frac{v}{m_i}\right)\operatorname{tr}\{(P_iC_{id}P_i')^{-1}\}, \tag{6.6.17}$$

and invoking the arithmetic mean–harmonic mean inequality,

$$\operatorname{tr}(\mathcal{I}_{sd}^{-1}) \geq \sum_{i=1}^{2}\left(\frac{v}{m_i}\right)(m_i - 1)^2\{\operatorname{tr}(P_iC_{id}P_i')\}^{-1}. \tag{6.6.18}$$

Note that C_{1d} and C_{2d} are like the information matrices for the block and varietal effects in a varietal block design, and we proceed as in a design of that kind to complete the proof (see Dey, 1986, ch. 2). By (6.6.12) and (6.6.16), for $i = 1, 2$, $C_{id}\mathbf{1}_{m_i} = \mathbf{0}$. Hence using (6.6.13) and (6.6.16),

$$\begin{aligned}\operatorname{tr}(P_1 C_{1d} P_1') &= \operatorname{tr}(C_{1d} P_1' P_1)\\ &= \operatorname{tr}(C_{1d})\\ &= N - \sum_{j_1=0}^{m_1-1}\sum_{j_2=0}^{m_2-1} \frac{\{b_d(j_1, j_2)\}^2}{r_{2d}(j_2)},\end{aligned} \tag{6.6.19}$$

where $b_d(j_1, j_2)$ is the (j_1, j_2)th element of B_d and $r_{2d}(j_2)$ is the j_2th element of $\boldsymbol{r}_{2d}$. Since the elements of B_d are integers, by (6.6.12) for $0 \le j_2 \le m_2 - 1$, we have

$$\sum_{j_1=0}^{m_1-1} \{b_d(j_1, j_2)\}^2 \ge \sum_{j_1=0}^{m_1-1} b_d(j_1, j_2) = r_{2d}(j_2).$$

From (6.6.19), recalling that $N = m_1 + m_2 - 1$, it now follows that

$$\operatorname{tr}(P_1 C_{1d} P_1') \le N - m_2 = m_1 - 1.$$

Similarly $\operatorname{tr}(P_2 C_{2d} P_2') \le m_2 - 1$. Therefore, by (6.6.18), for any Resolution (1, 1) plan $d \in \mathcal{D}_N$,

$$\operatorname{tr}(\mathcal{I}_{sd}{}^{-1}) \ge \sum_{i=1}^{2} \left(\frac{v}{m_i}\right)(m_i - 1). \tag{6.6.20}$$

Consider now the plan d_0 as specified in the statement of the theorem. By (6.6.16),

$$C_{id_0} = I_{m_i} - m_i{}^{-1}\mathbf{1}_{m_i}\mathbf{1}'_{m_i},$$

so $P_i C_{id_0} P_i' = I_{m_i - 1}$, $i = 1, 2$. Hence by (6.6.9), (6.6.14) and (6.6.15), $\mathcal{I}_{sd_0}$ is p.d., while, in consideration of (6.6.17), equality holds in (6.6.20) for d_0. Thus the claimed A_s-optimality of d_0 follows. ■

Remark 6.6.1. **(a)** Unlike what happens under the D- (or, equivalently, under the D_s-) criterion, not all saturated Resolution (1, 1) plans are equivalent under the A_s-criterion in the two factor case. For example, if $m_1 = 2$, $m_2 = 3$, then the saturated Resolution (1, 1) plan $d_1 = \{00, 01, 11, 12\}$ is inferior to d_0 of Theorem 6.6.2 with regard to the A_s-criterion.

(b) In the setup of Theorem 6.6.2, Mukerjee et al. (1986) claimed the optimality of d_0 with respect to a criterion similar to the E_s-criterion. However, there was a gap in their proof which was subsequently rectified by Krafft and Schaefer (1991).

(c) It is possible to develop an alternative proof of Theorem 6.6.1 using graph-theoretic arguments. We refer to Bapat and Dey (1991) for such an approach in the context of a corresponding problem in varietal block designs. ■

Even with $n = 2$ factors, the optimal design problem gets considerably harder in the unsaturated case $N > m_1 + m_2 - 1$. However, some progress has been made in the nearly saturated situation $N = m_1 + m_2$. Two of these results are summarized below.

Theorem 6.6.3. *Let $n = 2$, $N = m_1 + m_2$, and without loss of generality, suppose that $m_1 \leq m_2$. For $i = 1, 2$, let $m_i{}' = m_i - 1$.*

(a) (Mukerjee and Sinha, 1990) The plan

$$d^* = \{00, 01, 11, 12, \ldots, m_1{}'m_1{}', m_1{}'m_1, \ldots, m_1{}'m_2{}', m_1{}'0\}$$

is a Resolution (1, 1) plan that is D-optimal (and hence D_s-optimal) in $\mathcal{D}_N$.

(b) (Birkes and Dodge, 1991) Consider a plan $\tilde{d}(\in \mathcal{D}_N)$ obtained by the augmentation of the run 11 to the plan d_0 of Theorem 6.6.2. If $m_1 \neq 3$, then $\tilde{d}$ is a Resolution (1, 1) plan that is A_s-optimal in $\mathcal{D}_N$. ■

Further discussion on the case $m_1 = 3$, not covered in Theorem 6.6.3(b), is available in Birkes and Dodge (1991). However, their normalization of main effect contrasts is slightly different from ours and, therefore, not all of their A_s-optimality results for $m_1 = 3$ are guaranteed to be valid in our setting. Birkes and Dodge (1991) also investigated the optimality of the plan $\tilde{d}$ of Theorem 6.6.3 (b) under an E_s-type criterion. Further results on $\tilde{d}$, with reference to generalized criteria of type 1, were reported by Collombier (1992). While Mukerjee and Sinha's (1990) proof of Theorem 6.6.3(a) involved matrix manipulation, it is possible to develop an alternative proof using graph-theoretic arguments. The interested reader is referred to Balasubramanian and Dey (1996) and Dey et al. (1995) for results based on such an approach in the context of a parallel problem in varietal block designs.

Turning to the three factor saturated case, we have the following result:

Theorem 6.6.4. *(Chatterjee and Mukerjee, 1993). Let $n = 3$ and $m_1 = 2 \leq m_2 \leq m_3$. Consider the saturated case $N = m_1 + m_2 + m_3 - 2 = m_2 + m_3$.*

Let $m_i' = m_i - 1$, $m_i'' = m_i - 2$ $(i = 2, 3)$. *Then the plan*

$$d^* = \{00m_3', 010, 021, \ldots, 0m_2'm_2'',$$
$$100, 111, \ldots, 1m_2'm_2', 1m_2'm_2, \ldots, 1m_2'm_3'\}$$

is a Resolution (1, 1) *plan that is D-optimal in* $\mathcal{D}_N$. ■

A counterpart of Theorem 6.6.4, applicable to general $m_1 \times m_2 \times m_3$ factorials, is as yet unknown. However, for saturated symmetric m^3 factorials, Pesotan and Raktoe (1988) derived a D-optimality result over a subclass of $\mathcal{D}_N$.

Remark 6.6.2. The comment made in Remark 6.6.1(a) about the nonequivalence of all Resolution (1, 1) plans holds also in the context of Theorems 6.6.3 and 6.6.4. However, in either of these theorems as well as Theorem 6.6.2, plans obtained from the stated optimal ones via a renaming of the levels of one or more factor(s) are also optimal in the same sense. ■

6.7 SOME OTHER PLANS

Before concluding this chapter, we briefly review some plans that, though not guaranteed to be optimal in the class of *all* plans having the same number of runs, are of interest from other considerations. The foremost among these are the proportional frequency plans and the plans based on balanced arrays. The issue of optimality is in general rather deep, and notwithstanding the results presented so far, one may encounter experimental situations where $N, n, m_1, \ldots, m_n$ and f are such that optimal plans, even under specific criteria, are not yet known. The plans discussed in this section may be worth consideration in such situations.

The idea of proportional frequency plans appears to have originated from Plackett (1946), though it was Addelman (1962a) who initiated a systematic study of such plans. Consider the setup of an $m_1 \times m_2 \times \cdots \times m_n$ factorial, and let $f = 1$. Then a $d \in \mathcal{D}_N$ is called a proportional frequency (PF) plan if

$$B_d^{(i,u)} = N^{-1}\boldsymbol{r}_{id}\boldsymbol{r}_{ud}', \quad 1 \le i, u \le n,\ i \ne u, \tag{6.7.1}$$

and

$$\boldsymbol{r}_{id} > \boldsymbol{0}, \quad 1 \le i \le n, \tag{6.7.2}$$

where the matrices $B_d^{(i,u)}$ and the vectors $\boldsymbol{r}_{id}$ are as defined in Section 6.4 and the strict inequality in (6.7.2) is to be interpreted elementwise. By (6.7.1), the frequency of occurrence of any level combination of any pair of distinct factors in a PF plan is proportional to the frequencies with which the corresponding individual levels occur therein.

As in the last section, let

$$\boldsymbol{\beta}_s^{(1)} = (\boldsymbol{\beta}_{10\ldots0}', \boldsymbol{\beta}_{01\ldots0}', \ldots, \boldsymbol{\beta}_{00\ldots1}')'$$

represent complete sets of orthonormal contrasts belonging to the main effects and $\mathcal{I}_{sd}$ be the information matrix for $\boldsymbol{\beta}_s^{(1)}$. The definition of $\mathcal{I}_{sd}$, given in (6.6.6) for $n = 2$, can be easily adapted for general n. Analogously to (6.6.9), it follows from (6.7.1) that for a PF plan d,

$$\begin{aligned}\mathcal{I}_{sd} = v^{-1}\text{diag}[&m_1 P_1(R_{1d} - N^{-1}\boldsymbol{r}_{1d}\boldsymbol{r}_{1d}')P_1', \ldots, \\ &m_n P_n(R_{nd} - N^{-1}\boldsymbol{r}_{nd}\boldsymbol{r}_{nd}')P_n'],\end{aligned} \tag{6.7.3}$$

where, as before, R_{id} is a diagonal matrix with diagonal entries given by the elements of $\boldsymbol{r}_{id}(1 \leq i \leq n)$. By (2.2.2), (6.7.2), and (6.7.3), $\mathcal{I}_{sd}$ is positive definite. Hence so is $\mathcal{I}_d$ (see (6.6.8)), which shows that a PF plan, as defined above, is of Resolution (1, 1). Furthermore, just as in Theorem 2.3.1, under d,

$$\begin{aligned}\mathbb{D}(\hat{\boldsymbol{\beta}}_s^{(1)}) &= \sigma^2 \mathcal{I}_{sd}^{-1} \\ &= \sigma^2 v \text{ diag } (m_1^{-1} P_1 R_{1d}^{-1} P_1', \ldots, m_n^{-1} P_n R_{nd}^{-1} P_n'),\end{aligned} \tag{6.7.4}$$

using (2.2.2) and (6.7.3). In view of the block-diagonal structure of (6.7.4), it is clear that if d is a PF plan, then the best linear unbiased estimators of contrasts belonging to *different* main effects are uncorrelated in d. It is not hard to see that the converse of this result is also true (e.g., see Addelman, 1962a). For this reason PF plans, as considered here for $f = 1$, have also been called orthogonal main effect plans in the literature.

While orthogonality in the sense of (6.7.4) might be intuitively appealing, examples can be given to demonstrate that PF plans are not necessarily optimal, even under specific criteria, in the relevant classes; see Srivastava and Ghosh (1996). On the other hand, by (6.7.1), a PF plan reduces to one given by an orthogonal array of strength two if, for each i, the elements of $\boldsymbol{r}_{id}$ are equal, in which case the plan is indeed universally optimal in $\mathcal{D}_N$ by Theorem 2.6.1. This suggests that a PF plan should behave well under the standard criteria if, for each i, the elements of $\boldsymbol{r}_{id}$ are as nearly equal as possible. For example, under the D-criterion, the point just mentioned becomes evident if from (6.7.3) one notes that $\det(\mathcal{I}_{sd})$ (or equivalently, $\det(\mathcal{I}_d)$ by (6.6.8)) is proportional to $\prod_{i=1}^{n} \det(R_{id})$ in a PF plan d. We refer to Adhikary and Das (1990) for more details in this regard.

Addelman (1962a) suggested a technique of collapsing to obtain PF plans from orthogonal arrays. Suppose that $m_i \leq m_i'(1 \leq i \leq n)$, and let an $OA(N, n, m_1' \times \cdots \times m_n', 2)$ be available. Then the technique of collapsing maps the m_i' symbols in the ith column of the orthogonal array, for each i, to m_i distinct symbols using a possibly many-to-one mapping. The resulting array, with rows interpreted as treatment combinations, is easily seen to represent a PF plan. For example, starting with a symmetric $OA(16, 5, 4, 2)$, shown in Example 4.2.1, if one replaces the symbols in the first column according to the scheme

$$0 \to 0, \quad 1 \to 1, \qquad 2,\ 3 \to 2,$$

then one gets a PF plan for a 3×4^4 factorial in 16 runs. We refer to Jacroux (1992, 1993), Street (1994), Burgess and Street (1994), and Street and Burgess (1994) for some recent results on PF plans. The first three of these papers deal with PF plans with a minimum number of runs while the last one is expository in nature.

It is straightforward to extend the notion of PF plans to the case of general f (see Addelman, 1962b). Mukerjee and Chatterjee (1985) noted that the resulting plans are of Resolution (f, f). However, as argued by Lewis and John (1976) and Mukerjee (1980), these plans are not even orthogonal for $f > 1$, while their optimality continues to remain doubtful. For this reason the details of such extension are not discussed here.

We next consider another important class of plans, namely those given by balanced arrays. Following Chakravarti (1956), a balanced array $BA(N, n, m, g)$, having N rows, $n(\geq 2)$ columns, m symbols, and strength $g(\leq n)$ is an $N \times n$ array with elements from a set of m distinct symbols, say $0, 1, \ldots, m-1$, in which each g-tuple $i_1 \ldots i_g$ occurs $\lambda_{i_1 \ldots i_g}$ times as a row of every $N \times g$ subarray $(0 \leq i_j \leq m-1,\ 1 \leq j \leq g)$, the nonnegative integer $\lambda_{i_1 \ldots i_g}$ being invariant under permutation of $i_1, \ldots, i_g$. In fact Chakravarti (1956) originally described an array of this kind as "partially balanced." The nomenclature "balanced array" is due to Srivastava and Chopra (1971a).

It is evident from the definition that the parameters of a balanced array remain invariant under permutation of columns. A balanced array reduces to an orthogonal array if the parameters $\lambda_{i_1 \ldots i_g}$ are all identical. In general, however, the $\lambda_{i_1 \ldots i_g}$'s are not all equal, and because of the resulting flexibility a $BA(N, n, m, g)$ is often available even when an $OA(N, n, m, g)$ is nonexistent. Two examples of balanced arrays are shown below:

$$BA(7,5,2,2) = \begin{bmatrix} 0 & 1 & 0 & 1 & 0 \\ 1 & 0 & 0 & 1 & 1 \\ 0 & 0 & 1 & 1 & 0 \\ 1 & 1 & 1 & 0 & 0 \\ 0 & 1 & 0 & 0 & 1 \\ 1 & 0 & 0 & 0 & 0 \\ 0 & 0 & 1 & 0 & 1 \end{bmatrix}, \tag{6.7.5}$$

where $\lambda_{00} = \lambda_{01} = \lambda_{10} = 2$, $\lambda_{11} = 1$.

$$BA(10,4,3,2) = \begin{bmatrix} 000 & 111 & 222 & 0 \\ 012 & 012 & 012 & 0 \\ 012 & 120 & 201 & 0 \\ 012 & 201 & 120 & 0 \end{bmatrix}', \tag{6.7.6}$$

where $\lambda_{00} = 2$ and $\lambda_{i_1 i_2} = 1$ whenever $i_1 i_2 \neq 00$.

Just as with an orthogonal array, the rows of a $BA(N, n, m, g)$ can be identified with the treatment combinations of a symmetric m^n factorial. Then the array represents an N-run plan for such a factorial. The plans arising from balanced arrays of even strength $g(= 2f)$ have received particular attention, and their status as possible Resolution (f, f) plans has been explored extensively. Because of the invariance of balanced arrays under permutation of columns, the information matrices of the corresponding plans, as defined in (6.1.1), remain invariant under renaming of factors. This property, in fact, characterizes balanced arrays—see Srivastava (1970) and Yamamoto et al. (1975) for details.

Given N, n, m, and f, the key issues in connection with plans based on balanced arrays are as follows:

(a) Does there exist a $BA(N, n, m, 2f)$, with suitably chosen parameters $\lambda_{i_1 \ldots i_{2f}}$, that yields a Resolution (f, f) plan, that is, keeps the associated information matrix positive definite?

(b) If the answer to (a) above is in the affirmative, then what is the best choice of the parameters $\lambda_{i_1 \ldots i_{2f}}$ with regard to the performance of the resulting plan under specific optimality criteria, like the A- or D-criteria?

To answer these questions, one needs to express the characteristic polynomial of the information matrix $\mathcal{I}_d$ of a plan d, arising from a balanced array, in terms of the parameters $\lambda_{i_1 \ldots i_{2f}}$ and then find algebraic expressions for $\mathrm{tr}(\mathcal{I}_d{}^{-1})$ or $\det(\mathcal{I}_d{}^{-1})$ in terms of these parameters when $\mathcal{I}_d$ is positive definite. Finally the parameters $\lambda_{i_1 \ldots i_{2f}}$ have to be chosen so as to minimize $\mathrm{tr}(\mathcal{I}_d{}^{-1})$ or $\det(\mathcal{I}_d{}^{-1})$. The underlying algebra is nontrivial, and we refer to Srivastava (1970) and Srivastava and Chopra (1971a, b) for pioneering results. The work of these authors was followed up by Yamamoto et al. (1976), Shirakura (1976) and others, and an informative account of the early developments is available in Srivastava (1978).

While the approach indicated in the previous paragraph yields an optimal N-run plan within the class of plans given by balanced arrays in a given context, the issue of global optimality (i.e., optimality within the entire class $\mathcal{D}_N$) of such a plan remains uncertain. Several authors, such as Kuwada (1982) and Nguyen and Dey (1989), demonstrated that the best plan, derived through a suitable choice of the parameters $\lambda_{i_1 \ldots i_{2f}}$ of a balanced array under a given criterion, may not be globally optimal under the same criterion. On the other hand, it is satisfying to note that under some circumstances, consideration of balanced arrays can yield globally optimal plans. Thus, if an $OA(N, n, m, 2f)$ exists and a row $00 \ldots 0$ is added to it, then the resulting structure is easily seen to represent a balanced array which, by Theorem 6.2.2, yields an optimal Resolution (f, f) plan in the entire class $\mathcal{D}_N$ with respect to every generalized criterion of type 1. For example, the array shown in (6.7.6) is obtained via augmentation of a row 0000 to an $OA(9, 4, 3, 2)$ and hence represents an optimal plan in $\mathcal{D}_{10}$ in the sense just described. Again, it can be checked that the balanced array in (6.7.5) yields a Resolution (1, 1) plan with information matrix $v^{-1}(8I_6 - J_{66})$, where $v = 2^5$, and that a comparison with (6.5.3)–(6.5.5) reveals its D-optimality in the entire class $\mathcal{D}_7$.

Recent years have witnessed a further growth in the literature on balanced arrays and related topics. In this connection the reader may refer to Shirakura (1993), Hyodo and Kuwada (1994), Shirakura and Tong (1996), and Srivastava (1996) where more references can be found.

Many other fractional factorial plans have been proposed from various considerations; see Dey (1985) for a fairly comprehensive list of references up to that stage. Several authors, like Saha et al. (1982), Pesotan and Raktoe (1985), Krehbiel and Anderson (1991), and Raghavarao et al. (1991), proposed plans or augmentations thereof that enjoy optimality properties over appropriate subclasses of competing designs though they are not guaranteed to be globally optimal. Anderson and Federer (1995) extended the work of Raktoe and Federer (1970) to general $m_1 \times \cdots \times m_n$ factorials and reported a formulation for the problem of D-optimality in terms of matrices with elements 0 and 1. In a slightly different context, John et al. (1995) studied minimax distance designs that do not presume any underlying model but still compare favorably with designs dictated by standard optimality criteria. Computer-aided construction of satisfactory plans under various optimality criteria has also been of recent interest; see in this context Nguyen and Dey (1989), Nguyen (1996b), Nguyen and Miller (1997), and the references therein.

EXERCISES

6.1. Verify (6.2.15)–(6.2.17) in full detail.

6.2. For $f = 1$, find the scalars $a(\boldsymbol{x})$ in (6.2.19) explicitly, and check that these are nonnegative.

6.3. In the spirit of Example 6.3.1, construct another example to demonstrate that under the D- or A-criteria the behaviour of plans obtained via augmentation of an orthogonal array by two runs can depend on the particular runs added.

6.4. For a 2^8 factorial, indicate the construction of a Resolution (2, 2) plan in 66 runs such that the plan is D- and A-optimal within the class of all plans involving the same number of runs. [*Hint*: Start with Example 3.4.3(b).]

6.5. In the spirit of Example 6.4.1, illustrate two more applications of Theorem 6.4.1.

6.6. Give an example to demonstrate that not all plans involving $N = m_1 + m_2$ runs are equivalent under the D-criterion in the setup of Theorem 6.6.3.

6.7. Give an example to show that a proportional frequency plan is not necessarily optimal under the D-, A-, or E-criteria over the class of plans involving the same number of runs.

CHAPTER 7

Trend-Free Plans and Blocking

7.1 INTRODUCTION

So far in this book we have not concerned ourselves with the order of the treatment combinations or runs within a fractional factorial plan. However, there are practical situations where such order must be duly taken into account. The most important situation of this kind is one where the experiment is carried over a period of time and the consecutive observations, under any given plan, are influenced by a time trend in addition to the factorial effects. Under the framework of a time trend that can be represented by a polynomial, we address the issue of optimality in Section 7.2 and, in this connection, introduce trend-free plans and the associated orthogonal arrays. The problem of constructing such arrays is taken up in Section 7.3 where we also indicate briefly the related issue of minimizing the cost of factor level changes. Finally in Section 7.4 we consider situations, where the experimental units are not homogeneous but can be grouped into homogeneous classes known as blocks, and explore the relevant fractional factorial plans.

7.2 TREND-FREE PLANS: BASIC PRINCIPLES

With reference to an $m_1 \times \cdots \times m_n$ factorial, suppose that interest lies in the general mean and factorial effects involving at most f factors under the absence of factorial effects involving $t+1$ or more factors $(1 \leq f \leq t \leq n-1)$. For any $d \in \mathcal{D}_N$, the class of N-run plans, let the successive observations correspond to equispaced points of time, and suppose that these observations are influenced by a time trend that can be represented by a polynomial of degree $s(1 \leq s \leq N-1)$. The linear model (2.3.12), introduced in Chapter 2, then needs to be revised in order to incorporate such a time trend.

Let $q_0(\cdot) \equiv 1, q_1(\cdot), \ldots, q_s(\cdot)$ be orthogonal polynomials of degrees $1, \ldots, s$, respectively, defined on the basis of equispaced time points, say $1, \ldots, N$. Then

$$\sum_{u=1}^{N} q_j(u) q_{j'}(u) = 0, \qquad 0 \le j \ne j' \le s.$$

For example, one can take

$$q_1(u) = u - \tfrac{1}{2}(N+1), \quad q_2(u) = u^2 - (N+1)u + \tfrac{1}{6}(N+1)(N+2), \quad (7.2.1)$$

and so on. As before, for positive integral a, let $\mathbf{1}_a$ be the $a \times 1$ vector with all elements unity and I_a be the identity matrix of order a. Also let $v = \prod m_i$.

In the spirit of Coster and Cheng (1988) and Lin and Dean (1991) among others, we represent the effect of trend at the time point u by $\sum_{k=1}^{s} \theta_k q_k(u)$, $1 \le u \le N$, where $\theta_1, \ldots, \theta_s$ are the trend parameters. Then the model (2.3.12), associated with the plan d, may be modified as

$$\mathbb{E}(\boldsymbol{Y}) = X_{1d}\boldsymbol{\beta}^{(1)} + X_{2d}\boldsymbol{\beta}^{(2)} + X_3\boldsymbol{\theta}, \quad \mathbb{D}(\boldsymbol{Y}) = \sigma^2 I_N. \qquad (7.2.2)$$

In the above, $\boldsymbol{Y} = (Y_1, \ldots, Y_N)'$ is the observational vector, with Y_u denoting the observation at the time point u and X_3 is an $N \times s$ matrix with (u, k)th element $q_k(u)$ $(1 \le u \le N; 1 \le k \le s)$. Furthermore $\boldsymbol{\theta} = (\theta_1, \ldots, \theta_s)'$, $\sigma^2(> 0)$ is the common variance of the observations, and as in (2.3.13),

$$X_{1d} = X_d(P^{(1)})', \qquad X_{2d} = X_d(P^{(2)})', \qquad (7.2.3)$$

the matrices X_d, $P^{(1)}$, $P^{(2)}$, and the parametric vectors $\boldsymbol{\beta}^{(1)}$, $\boldsymbol{\beta}^{(2)}$ being as specified in (2.3.2) and (2.3.8)–(2.3.10). It will be convenient to note here that X_d is the $N \times v$ design matrix corresponding to d, with rows indexed by u and columns indexed by the lexicographically ordered treatment combinations, such that the $(u, j_1 \ldots j_n)$th element of X_d is $\chi_d(u; j_1 \ldots j_n)$, where the indicator $\chi_d(u; j_1 \ldots j_n)$ equals 1 if the observation at time point u according to d is associated with the treatment combination $j_1 \ldots j_n$ and zero otherwise. Also recall that $\boldsymbol{\beta}^{(1)}$, representing the general mean and effects involving at most f factors, is the parametric vector of interest. In particular, since $P^{00\ldots0} = v^{-1/2}\mathbf{1}'_v$ (see (2.2.4)), it is clear from (2.3.8), (2.3.9), (7.2.2), and (7.2.3) that the element $\beta_{00\ldots0}$ of $\boldsymbol{\beta}^{(1)}$, which corresponds to the general mean, contributes a term $v^{-1/2}\beta_{00\ldots0}\mathbf{1}_N$ to $\mathbb{E}(\boldsymbol{Y})$. At this stage it may be remarked that we are not including $q_0(\cdot)(\equiv 1)$ in the trend effect, since this amounts to the addition of a constant to the expectation of each observation and, as just noted, $\beta_{00\ldots0}$ already does the job.

Let $\mathcal{I}_d^{\text{trend}}$ denote the information matrix for $\boldsymbol{\beta}^{(1)}$ under the plan d and the model (7.2.2). Analogously to (2.3.14),

$$\mathcal{I}_d^{\text{trend}} = X_{1d}'X_{1d} - X_{1d}'(X_{2d}X_3)\begin{pmatrix} X_{2d}'X_{2d} & X_{2d}'X_3 \\ X_3'X_{2d} & X_3'X_3 \end{pmatrix}^{-}\begin{pmatrix} X_{2d}' \\ X_3' \end{pmatrix} X_{1d}, \quad \text{if } f < t, \qquad (7.2.4)$$

$$= X_{1d}'X_{1d} - X_{1d}'X_3(X_3'X_3)^{-}X_3'X_{1d}, \qquad \text{if } f = t. \qquad (7.2.5)$$

Exactly as in Theorem 2.3.1, under the model (7.2.2), $\boldsymbol{\beta}^{(1)}$ is estimable in d if and only if $\mathcal{I}_d^{\text{trend}}$ is positive definite, in which case we will continue to call d a Resolution (f, t) plan.

We now turn to the issue of optimality in the presence of a time trend. To that effect, first note that for any $d \in \mathcal{D}_N$, $\mathcal{I}_d - \mathcal{I}_d^{\text{trend}}$ is nonnegative definite where $\mathcal{I}_d$ is the information matrix for $\boldsymbol{\beta}^{(1)}$ under d and the model (2.3.12) arising when there is no time trend. Recall that (2.3.14) gives an expression for $\mathcal{I}_d$ and that the second term in the right-hand side of (2.3.14) does not occur when $f = t$. With reference to the model (7.2.2), a plan $d(\in \mathcal{D}_N)$ will be called s-trend-free or s-trend-resistant if $\mathcal{I}_d^{\text{trend}}$ equals $\mathcal{I}_d$. For such a plan the presence of the trend parameters $\theta_1, \ldots, \theta_s$ in the model entails no loss of information on $\boldsymbol{\beta}^{(1)}$. Hence one may anticipate a key role for s-trend-free plans in the context of optimality and wish to characterize the underlying conditions. This is done in Theorem 7.2.1 below for plans based on orthogonal arrays. In the sequel, for any $d(\in \mathcal{D}_N)$, any $u(1 \le u \le N)$, and any $i_1, \ldots, i_f, j_{i_1}, \ldots, j_{i_f}(1 \le i_1 < \cdots < i_f \le n; 0 \le j_{i_1} \le m_{i_1} - 1, \ldots, 0 \le j_{i_f} \le m_{i_f} - 1)$, the indicator $\chi_d^{i_1 \ldots i_f}(u; j_{i_1} \ldots j_{i_f})$ assumes the value 1 if the observation at the time point u according to d corresponds to levels $j_{i_1}, \ldots, j_{i_f}$ of the i_1th, $\ldots$, i_fth factors, respectively, and the value zero otherwise. Clearly,

$$\chi_d^{i_1 \ldots i_f}(u; j_{i_1} \ldots j_{i_f}) = \sum \cdots \sum \chi_d(u; j_1 \ldots j_n), \qquad (7.2.6)$$

where the summation in the right-hand side extends over $j_i (i \ne i_1, \ldots, i_f)$.

Theorem 7.2.1. *Let $d_0(\in \mathcal{D}_N)$ be represented by an orthogonal array $OA(N, n, m_1 \times \cdots \times m_n, f + t)$.*

(a) *Then d_0 is s-trend-free if and only if*

$$\sum_{u=1}^{N} u^k \chi_{d_0}^{i_1 \ldots i_f}(u; j_{i_1} \ldots j_{i_f}) = (m_{i_1} \cdots m_{i_f})^{-1} \sum_{u=1}^{N} u^k, \quad 1 \le k \le s, \qquad (7.2.7)$$

for every $i_1, \ldots, i_f$ *and* $j_{i_1}, \ldots, j_{i_f}$ $(1 \leq i_1 < \cdots < i_f \leq n;\ 0 \leq j_{i_1} \leq m_{i_1} - 1, \ldots, 0 \leq j_{i_f} \leq m_{i_f} - 1)$.

(b) *Furthermore, under (7.2.7),* d_0 *is a universally optimal Resolution* (f, t) *plan in* $\mathcal{D}_N$.

Proof.

(a) First suppose that $f < t$. Then by (7.2.3),

$$X_{1d_0}{}'X_{2d_0} = P^{(1)}R_{d_0}\left(P^{(2)}\right)', \tag{7.2.8}$$

where, as before, $R_{d_0} = X_{d_0}{}'X_{d_0}$ is a $v \times v$ diagonal matrix with diagonal elements given by the frequencies with which the v treatment combinations appear in d_0. Since d_0 is represented by an $OA(N, n, m_1 \times \cdots \times m_n, f + t)$, by (7.2.8) and Lemma 2.6.1(b), $X_{1d_0}{}'X_{2d_0} = O$. Therefore, in consideration of (2.3.14) and (7.2.4), d_0 is s-trend-free if and only if

$$X_{1d_0}{}'X_3 = O. \tag{7.2.9}$$

Trivially, by (7.2.5), the condition (7.2.9) is necessary and sufficient for d_0 to be s-trend-free also when $f = t$. Thus we need to show the equivalence of (7.2.7) and (7.2.9). In order to keep the notation simple, this will be done for $f = 2$. The treatment is similar for general f and even simpler for $f = 1$.

Let (7.2.9) hold. For $1 \leq k \leq s$, let $\boldsymbol{\xi}_k$ be a $v \times 1$ vector with elements

$$\sum_{u=1}^{N} q_k(u)\chi_{d_0}(u;\ j_1 \ldots j_n),$$

arranged in the lexicographic order. Under (7.2.9), recalling the definitions of X_{d_0} and X_3, it follows from (7.2.3) that

$$P^{(1)}\boldsymbol{\xi}_k = \mathbf{0}, \qquad 1 \leq k \leq s. \tag{7.2.10}$$

As $f = 2$, from (2.3.9) and (7.2.10) one gets

$$\begin{bmatrix} P^{00\ldots0} \\ P^{10\ldots0} \\ P^{01\ldots0} \\ P^{11\ldots0} \end{bmatrix} \boldsymbol{\xi}_k = \mathbf{0}, \qquad 1 \leq k \leq s. \tag{7.2.11}$$

However, by (2.2.4) and (2.2.5),

$$\begin{bmatrix} P^{00\ldots0} \\ P^{10\ldots0} \\ P^{01\ldots0} \\ P^{11\ldots0} \end{bmatrix} = \sqrt{\frac{m_1 m_2}{v}} \begin{bmatrix} (m_1{}^{-1/2}\mathbf{1}'_{m_1}) \otimes (m_2{}^{-1/2}\mathbf{1}'_{m_2}) \\ P_1 \otimes (m_2{}^{-1/2}\mathbf{1}'_{m_2}) \\ (m_1{}^{-1/2}\mathbf{1}'_{m_1}) \otimes P_2 \\ P_1 \otimes P_2 \end{bmatrix}$$
$$\times [I_{m_1} \otimes I_{m_2} \otimes \mathbf{1}'_{m_3} \otimes \cdots \otimes \mathbf{1}'_{m_n}].$$

By (2.2.2), the first matrix appearing in the right-hand side of the above is orthogonal. Hence from (7.2.11),

$$[I_{m_1} \otimes I_{m_2} \otimes \mathbf{1}'_{m_3} \otimes \cdots \otimes \mathbf{1}'_{m_n}]\boldsymbol{\xi}_k = \mathbf{0}, \qquad 1 \le k \le s.$$

By (7.2.6) and the definition of $\boldsymbol{\xi}_k$, the above implies that

$$\sum_{u=1}^{N} q_k(u)\chi_{d_0}{}^{12}(u;\, j_1 j_2) = 0, \qquad 1 \le k \le s, \tag{7.2.12}$$

for every j_1, j_2 ($0 \le j_1 \le m_1 - 1, 0 \le j_2 \le m_2 - 1$). Since d_0 is represented by an orthogonal array of strength $f + t (> 2)$, we also have

$$\sum_{u=1}^{N} \chi_{d_0}{}^{12}(u;\, j_1 j_2) = \frac{N}{(m_1 m_2)},$$

for every j_1, j_2. Hence from (7.2.12) and expressions like those in (7.2.1), it is seen recursively that for $1 \le k \le s$,

$$\sum_{u=1}^{N} u^k \chi_{d_0}{}^{12}(u;\, j_1 j_2)$$

does not depend on j_1, j_2. Since

$$\sum_{j_1=0}^{m_1-1}\sum_{j_2=0}^{m_2-1}\sum_{u=1}^{N} u^k \chi_{d_0}{}^{12}(u;\, j_1 j_2) = \sum_{u=1}^{N} u^k \left\{\sum_{j_1=0}^{m_1-1}\sum_{j_2=0}^{m_2-1} \chi_{d_0}{}^{12}(u;\, j_1 j_2)\right\}$$
$$= \sum_{u=1}^{N} u^k,$$

it follows that

$$\sum_{u=1}^{N} u^k \chi_{d_0}{}^{12}(u;\, j_1 j_2) = (m_1 m_2)^{-1} \sum_{u=1}^{N} u^k, \qquad 1 \le k \le s,$$

for every j_1, j_2 $(0 \le j_1 \le m_1-1, 0 \le j_2 \le m_2-1)$. Thus (7.2.7) holds for $i_1 = 1$ and $i_2 = 2$. Similarly it can be seen that (7.2.9) implies (7.2.7) for every i_1, i_2 and j_{i_1}, j_{i_2} $(1 \le i_1 < i_2 \le n; 0 \le j_{i_1} \le m_{i_1} - 1, 0 \le j_{i_2} \le m_{i_2} - 1)$.

The converse can be proved by reversing the above steps.

(b) Follows as in the proof of Theorem 2.6.1 noting that for any $d \in \mathcal{D}_N$,

$$\mathrm{tr}(\mathcal{I}_d^{\mathrm{trend}}) \le \mathrm{tr}(\mathcal{I}_d),$$

because of the nonnegative definiteness of $\mathcal{I}_d - \mathcal{I}_d^{\mathrm{trend}}$, and that $\mathcal{I}_{d_0}^{\mathrm{trend}}$ equals $\mathcal{I}_{d_0}$ under (7.2.7). ■

Theorem 7.2.1, whose root can be traced to the work of Phillips (1968), is of considerable help in the study of optimal fractional factorial plans when the observations are influenced by a time trend. It shows that if the order of rows in an orthogonal array and hence the order of runs in the associated fractional factorial plan satisfy (7.2.7), then the universal optimality result proved earlier in Theorem 2.6.1 continues to remain valid even under a time trend. Incidentally, the equivalence of (7.2.7) and (7.2.9), noted while proving Theorem 7.2.1, shows that a plan d_0, as envisaged in this theorem, leads to estimation of the parametric vector of interest, namely $\boldsymbol{\beta}^{(1)}$, orthogonally to trend parameters.

Observe that (7.2.7) depends on both f and s but not explicitly on t. Hence an orthogonal array will be called trend-free of order $(f, s)(f, s \ge 1)$ if it has strength at least f and, in addition, the associated plan satisfies (7.2.7) for every $i_1, \ldots, i_f$ and $j_{i_1}, \ldots, j_{i_f}$. Using identities similar to (7.2.6), it is easy to see from (7.2.7) that such an array is also trend-free of order (f', s') for every choice of positive integers $f'(\le f)$ and $s'(\le s)$. Trivially, an orthogonal array is trend-free of order $(f, 0)(f \ge 1)$ if it has strength at least f but is not guaranteed to be trend-free of order (f, s) even for $s = 1$. The following corollary, which is another way of stating Theorem 7.2.1, is evident:

Corollary 7.2.1. *Let $d_0(\in \mathcal{D}_N)$ be represented by an $OA(N, n, m_1 \times \cdots \times m_n, g)$ which is trend-free of order (f, s), where $2 \le g \le n$, $1 \le f \le g$ and $s \ge 1$. Then for every choice of integers f', s' satisfying*

$$1 \le f' \le \min\left(\tfrac{1}{2}g, f\right), \qquad 1 \le s' \le s,$$

d_0 is a universally optimal Resolution $(f', g - f')$ plan in $\mathcal{D}_N$ in the presence of a polynomial time trend of degree s'. ■

The problem of constructing trend-free orthogonal arrays, which has received much attention in the literature, will be taken up in the next section. However, two illustrative examples are presented below.

Example 7.2.1. Let N be even, and suppose that an $OA(\frac{1}{2}N, n, m_1 \times \cdots \times m_n, g)$, say A_1, of strength $g(\geq 2)$ exists. Let A_2 be an array obtained from A_1 by arranging the rows of A_1 in the reversed order. Then it is easy to see that $A = [A_1{}'A_2{}']'$ is an $OA(N, n, m_1 \times \cdots \times m_n, g)$ which is trend-free of order $(g, 1)$. For every f' satisfying $1 \leq f' \leq \frac{1}{2}g$, by Corollary 7.2.1, A represents a universally optimal Resolution $(f', g - f')$ plan in $\mathcal{D}_N$ in the presence of a linear trend. This method of construction is, however, not very efficient, since it doubles the number of runs to take care of only one parameter corresponding to a linear trend. ■

Example 7.2.2. (Jacroux and Saharay, 1991). Let there exist Hadamard matrices H_{N_1} and H_{N_2}, of orders N_1 and N_2, respectively, where $N_1 \geq 2$ and $N_2 \geq 4$. For $i = 1, 2$, without loss of generality, suppose that the first column of H_{N_i} consists of only $+1$'s, and let $H_{N_i}{}^*$ be an $N_i \times (N_i - 1)$ matrix obtained by deleting the first column of H_{N_i}. Then by (7.2.7), the array A obtained upon replacement of each -1 in $H_{N_1}{}^* \otimes H_{N_2}{}^*$ by zero is a symmetric $OA(N_1N_2, (N_1 - 1)(N_2 - 1), 2, 2)$ which is trend-free of order $(1, 1)$. As before, A represents a universally optimal Resolution $(1, 1)$ plan in $\mathcal{D}_N$, with $N = N_1N_2$, in the presence of a linear trend. ■

Remark 7.2.1. Although Theorem 7.2.1 plays a crucial role in the study of optimality under the presence of trend, there are situations where this theorem is not applicable. First of all, the values of $N, n, m_1, \ldots, m_n, f$, and t in a given context may be such that an $OA(N, n, m_1 \times \cdots \times m_n, f + t)$ is nonexistent. Second, even if such an array exists, s may be such that no run order satisfying (7.2.7) is available. Occasionally this second difficulty may be obvious noting that the quantities in the left-hand side and hence those in the right-hand side of (7.2.7) must be integers whenever (7.2.7) holds. For example, if $N = 18$, $m_1 = \cdots = m_{n-1} = 3$, $m_n = 6$, $f = t = 1$, and $s = 1$, then an $OA(18, n, 3^{n-1} \times 6, 2)$ exists provided that $2 \leq n \leq 7$ (see Example 4.5.2), but it is impossible to satisfy (7.2.7), since $\frac{1}{6}\sum_{u=1}^{18} u$ is not an integer. Furthermore Theorem 7.2.1 ceases to be meaningful if the successive observations correspond to nonequispaced points of time. The literature on optimal fractional factorial plans in the presence of trend, when Theorem 7.2.1 fails, is very scanty, and we refer to Atkinson and Donev (1996) for more discussion and illustrative examples. ■

7.3 TREND-FREE ORTHOGONAL ARRAYS

In this section we will be primarily concerned with the issue of construction of trend-free orthogonal arrays. Lemma 7.3.1 below, due to Jacroux (1994), will be of much help. In the sequel, an $N \times 1$ vector $\boldsymbol{a} = (a_1, \ldots, a_N)'$ with elements from a finite additive Abelian group $\mathcal{G} = \{\gamma_0, \gamma_1, \ldots, \gamma_{m-1}\}$ of order $m(\geq 2)$ will be called s-trend-free ($s \geq 0$), if analogously to (7.2.7),

$$\sum_{u=1}^{N} u^k \chi_{\boldsymbol{a}}(u; j) = m^{-1} \sum_{u=1}^{N} u^k, \qquad 0 \leq k \leq s, \quad 0 \leq j \leq m-1, \quad (7.3.1)$$

where

$$\chi_{\boldsymbol{a}}(u; j) = \begin{cases} 1 & \text{if } a_u = \gamma_j, \\ 0 & \text{otherwise.} \end{cases} \quad (7.3.2)$$

While, in the context of (7.2.7), d_0 is represented by an orthogonal array of strength $f + t$ so that (7.2.7) holds trivially for $k = 0$, the explicit inclusion of $k = 0$ in (7.3.1) guarantees that the members of $\mathcal{G}$ occur equally often as elements of $\boldsymbol{a}$, thereby implying that under (7.3.1) $\boldsymbol{a}$ is an $OA(N, 1, m, 1)$ which consists of a single column and is trend-free of order $(1, s)$. We will write $\odot$ for Kronecker sum, interpret γ_0 as the zero element of $\mathcal{G}$, and denote an $N \times 1$ vector with each element γ_0 by $\mathbf{0}_N$.

Lemma 7.3.1. *Let $\boldsymbol{a}_1$ and $\boldsymbol{a}_2$ be $N_1 \times 1$ and $N_2 \times 1$ vectors both with elements from the same finite additive Abelian group $\mathcal{G} = \{\gamma_0, \gamma_1, \ldots, \gamma_{m-1}\}$ of order $m(\geq 2)$. If $\boldsymbol{a}_j$ is s_j-trend-free ($s_j \geq 0$), $j = 1, 2$, then*

- ***(a)*** *$\boldsymbol{a}_1 \odot \mathbf{0}_{N_2}$ is s_1-trend-free;*
- ***(b)*** *$\mathbf{0}_{N_1} \odot \boldsymbol{a}_2$ is s_2-trend-free;*
- ***(c)*** *$\boldsymbol{a}_1 \odot \boldsymbol{a}_2$ is $(s_1 + s_2 + 1)$-trend-free.*

Proof. We prove only part (c); the proofs of parts (a) and (b) are similar and simpler. Consider any fixed $j (0 \leq j \leq m-1)$. For any $i (0 \leq i \leq m-1)$, there exists a unique $p_i (0 \leq p_i \leq m-1)$ such that

$$\gamma_i + \gamma_{p_i} = \gamma_j. \quad (7.3.3)$$

Note that $\{p_0, p_1, \ldots, p_{m-1}\}$ is a permutation of $\{0, 1, \ldots, m-1\}$. Let $\boldsymbol{a} = \boldsymbol{a}_1 \odot \boldsymbol{a}_2$. Then by (7.3.2), (7.3.3), and the definition of Kronecker sum (see Section 4.4), it follows that

$$\chi_{\boldsymbol{a}}(N_2(u_1 - 1) + u_2; j) = \sum_{i=0}^{m-1} \chi_{\boldsymbol{a}_1}(u_1; i) \chi_{\boldsymbol{a}_2}(u_2; p_i), \quad (7.3.4)$$

for every u_1, u_2 ($1 \leq u_1 \leq N_1$, $1 \leq u_2 \leq N_2$). Let $N = N_1 N_2$. Then, using (7.3.4), for $0 \leq k \leq s_1 + s_2 + 1$,

$$\sum_{u=1}^{N} u^k \chi_{\boldsymbol{a}}(u; j) = \sum_{u_1=1}^{N_1} \sum_{u_2=1}^{N_2} \{N_2(u_1 - 1) + u_2\}^k \chi_{\boldsymbol{a}}(N_2(u_1 - 1) + u_2; j)$$

$$= \sum_{l=0}^{k} \binom{k}{l} N_2{}^l \psi_l, \tag{7.3.5}$$

where

$$\psi_l = \sum_{i=0}^{m-1} \sum_{u_1=1}^{N_1} \sum_{u_2=1}^{N_2} (u_1 - 1)^l u_2^{k-l} \chi_{\boldsymbol{a}_1}(u_1; i) \chi_{\boldsymbol{a}_2}(u_2; p_i), \qquad 0 \leq l \leq k. \tag{7.3.6}$$

Since $k \leq s_1 + s_2 + 1$, for any integer l satisfying $0 \leq l \leq k$, either $l \leq s_1$ or $k - l \leq s_2$. First suppose that $l \leq s_1$. Then recalling that $\boldsymbol{a}_1$ is s_1-trend-free and using a binomial expansion for $(u_1 - 1)^l$, one gets

$$\sum_{u_1=1}^{N_1} (u_1 - 1)^l \chi_{\boldsymbol{a}_1}(u_1; i) = m^{-1} \sum_{u_1=1}^{N_1} (u_1 - 1)^l, \qquad 0 \leq i \leq m - 1. \tag{7.3.7}$$

Since $\{p_0, p_1, \ldots, p_{m-1}\}$ is a permutation of $\{0, 1, \ldots, m - 1\}$, we also have

$$\sum_{i=0}^{m-1} \chi_{\boldsymbol{a}_2}(u_2; p_i) = 1, \qquad 1 \leq u_2 \leq N_2. \tag{7.3.8}$$

From (7.3.6)–(7.3.8), it follows that if $l \leq s_1$, then

$$\psi_l = m^{-1} \sum_{u_1=1}^{N_1} \sum_{u_2=1}^{N_2} (u_1 - 1)^l u_2^{k-l}. \tag{7.3.9}$$

Next let $k - l \leq s_2$. Then recalling that $\boldsymbol{a}_2$ is s_2-trend-free,

$$\sum_{u_2=1}^{N_2} u_2^{k-l} \chi_{\boldsymbol{a}_2}(u_2; p_i) = m^{-1} \sum_{u_2=1}^{N_2} u_2^{k-l}, \qquad 0 \leq i \leq m - 1,$$

and as before, (7.3.9) is again seen to hold. Thus (7.3.9) holds for $0 \leq l \leq k$. From (7.3.5) and (7.3.9), it is clear that for $0 \leq k \leq s_1 + s_2 + 1$ and any fixed

$j(0 \leq j \leq m-1)$,

$$\sum_{u=1}^{N} u^k \chi_a(u; j) = m^{-1} \sum_{u_1=1}^{N_1} \sum_{u_2=1}^{N_2} \{N_2(u_1 - 1) + u_2\}^k = m^{-1} \sum_{u=1}^{N} u^k,$$

and part (c) of the lemma follows. ∎

Lemma 7.3.1 enables one to obtain trend-free arrays by suitably adapting some of the previously described methods of construction for orthogonal arrays. Considering, for instance, the foldover technique, discussed in Section 3.3, we get the following result due to John (1990):

Theorem 7.3.1. *The existence of a symmetric orthogonal array* $OA(N, n, 2, g)$ *that is trend-free of order* $(1, s)$ *implies that of an* $OA(2N, n, 2, g)$ *that is trend-free of order* $(1, s+1)$.

Proof. Let A be an $OA(N, n, 2, g)$ that is trend-free of order $(1, s)$. Denote the symbols in A by 0 and 1, and let $\bar{A}$ be the array obtained by interchanging the symbols 0 and 1 in A. Analogously to (3.3.1), define $A^* = [A' \quad \bar{A}']'$. Then A^* is an $OA(2N, n, 2, g)$. Note that with 0 and 1 interpreted as elements of the finite field $GF(2)$,

$$A^* = (0, 1)' \odot A. \tag{7.3.10}$$

Since A is trend-free of order $(1, s)$, by (7.3.1), each column of A is s-trend-free. Also, trivially, $(0, 1)'$ is 0-trend-free. Hence by (7.3.10) and Lemma 7.3.1(c), each column of A^* is $(s+1)$-trend-free, that is, A^* is trend-free of order $(1, s+1)$. ∎

Remark 7.3.1. In the setup of Theorem 7.3.1, if g is even, then, by Theorem 3.3.1, A^* is actually an $OA(2N, n, 2, g+1)$. Also the initial orthogonal array A is always trend-free of order $(1, 0)$. Hence it is always guaranteed that the foldover array A^* will be trend-free of order $(1, 1)$. From Theorems 3.2.2 and 7.3.1, it is now clear that the existence of a Hadamard matrix of order $N(\geq 4)$ implies that of an $OA(2N, N-1, 2, 3)$ which is trend-free of order $(1, 1)$. ∎

Remark 7.3.2. By generalizing Theorem 7.3.1, it can be seen that the existence of a symmetric $OA(N, n, m, g)$, say A, which is trend-free of order $(1, s)$ implies that of an $OA(mN, n, m, g)$ which is trend-free of order $(1, s+1)$. This follows if one denotes the symbols in A by the members of the group $\{0, 1, \ldots, m-1\}$, the group operation being addition modulo m, and then, analogously to (7.3.10), considers the array

$$A^* = (0, 1, \ldots, m-1)' \odot A. \qquad \blacksquare$$

Considering next the method of differences introduced in Section 3.5, we get the following result which strengthens an earlier one due to Chatterjee and Mukerjee (1986b):

Theorem 7.3.2. *For positive integers $\lambda(\geq 1)$, $m(\geq 2)$ and $n(\geq 3)$, suppose that there exists a difference matrix $D(\lambda m, n; m)$ with elements from a finite additive Abelian group $\mathcal{G}$ of order m.*

(a) *Then there exists a symmetric $OA(\lambda m^2, n-1, m, 2)$ that is trend-free of order $(1, 1)$.*

(b) *Furthermore one can construct a symmetric $OA(\lambda m^2, n, m, 2)$ that is trend-free of order $(1, 1)$, provided that either (i) λ is even or (ii) λ and m are both odd and $\lambda \geq 3$.*

Proof.

(a) Let $\mathcal{G} = \{\gamma_0, \gamma_1, \ldots, \gamma_{m-1}\}$ and as before, denote the zero element of $\mathcal{G}$ by γ_0. Since a difference matrix still remains so when the same member of $\mathcal{G}$ is added to each element of any row, without loss of generality, it may be assumed that every element in one of the columns of $D(\lambda m, n; m)$ equals γ_0. Let A_0 be a $\lambda m \times (n-1)$ matrix obtained from $D(\lambda m, n; m)$ by deleting such a column. For $1 \leq j \leq m-1$, let A_j be a $\lambda m \times (n-1)$ matrix obtained by adding γ_j to each element of A_0. Then, as in Theorem 3.5.1, the array

$$A^* = (A_0{}' A_1{}' \cdots A_{m-1}{}')'$$

is an $OA(\lambda m^2, n-1, m, 2)$. Note that

$$A^* = (\gamma_0, \gamma_1, \ldots, \gamma_{m-1})' \odot A_0. \qquad (7.3.11)$$

By construction, each member of $\mathcal{G}$ appears λ times in every column of A_0; that is, every column of A_0 is 0-trend-free. Hence, as with Theorem 7.3.1, it follows from (7.3.11) that the array A^* is trend-free of order $(1, 1)$.

(b) Let $\boldsymbol{a}_1 = (\gamma_0, \gamma_1, \ldots, \gamma_{m-1}, \gamma_{m-1}, \ldots, \gamma_1, \gamma_0)'$. Also, for odd m (e.g. $m = 2m_0 + 1$), define the $3m \times 1$ vector $\boldsymbol{a}_2 = (\boldsymbol{a}_{21}{}', \boldsymbol{a}_{22}{}', \boldsymbol{a}_{23}{}')'$, where

$$\boldsymbol{a}_{21} = (\gamma_0, \gamma_1, \ldots, \gamma_{m-1})',$$

$$\boldsymbol{a}_{22} = (\gamma_{m_0}, \gamma_{m_0+1}, \ldots, \gamma_{m-1}, \gamma_0, \gamma_1, \ldots, \gamma_{m_0-1})',$$

$$\boldsymbol{a}_{23} = (\gamma_{m-1}, \gamma_{m_0-1}, \gamma_{m-2}, \gamma_{m_0-2}, \ldots, \gamma_{m_0+1}, \gamma_0, \gamma_{m_0})'.$$

For example, if $m = 5$, then

$$\boldsymbol{a}_2 = (\gamma_0, \gamma_1, \gamma_2, \gamma_3, \gamma_4, \gamma_2, \gamma_3, \gamma_4, \gamma_0, \gamma_1, \gamma_4, \gamma_1, \gamma_3, \gamma_0, \gamma_2)'.$$

If (i) holds, then $\lambda(= 2\lambda_0$ say) is even, and we define

$$\boldsymbol{a} = (\boldsymbol{a}_1', \ldots, \boldsymbol{a}_1')', \tag{7.3.12}$$

$\boldsymbol{a}_1$ being repeated λ_0 times in the right-hand side of (7.3.12). On the other hand, if (ii) holds, then $\lambda(= 2\lambda_0 + 1$ say, $\lambda_0 \geq 1)$ and m are both odd, and we define

$$\boldsymbol{a} = (\boldsymbol{a}_2', \boldsymbol{a}_1', \ldots, \boldsymbol{a}_1')', \tag{7.3.13}$$

$\boldsymbol{a}_1$ being repeated $\lambda_0 - 1$ times in the right-hand side of (7.3.13). Under either (7.3.12) or (7.3.13), $\boldsymbol{a}$ is a $\lambda m \times 1$ vector that is 1-trend-free. Hence by Lemma 7.3.1(b), $\boldsymbol{0}_m \odot \boldsymbol{a}$ is also 1-trend-free. Consequently, with A^* defined as in part (a), it is easy to see that the $\lambda m^2 \times n$ array

$$[A^* \vdots \boldsymbol{0}_m \odot \boldsymbol{a}]$$

is an $OA(\lambda m^2, n, m, 2)$ that is trend-free of order (1, 1). ■

Remark 7.3.3. If λ is odd and m is even, then $m^{-1} \sum_{u=1}^{\lambda m} u = \frac{1}{2}\lambda(\lambda m + 1)$ is not an integer, and a comparison with (7.3.1) shows the nonexistence of a 1-trend-free vector, of order $\lambda m \times 1$, with elements from $\mathcal{G}$. Therefore, for odd λ and even m, the technique of adding an extra column, as employed while proving Theorem 7.3.2 (b), does not work. ■

Example 7.3.1. Consider the difference matrix $D(6, 6; 3)$ shown in (3.5.1). Its fourth column has each element equal to zero. One may obtain the matrix A_0 by deleting this column and then use (7.3.11) to get an $OA(18, 5, 3, 2)$ that is trend-free of order (1, 1). This array is the same as what one obtains by deleting the fourth and seventh columns of the array shown in Example 3.5.1. Here $m = 3$, $\lambda = 2$ is even, and following the proof of Theorem 7.3.2(b), augmentation of the column

$$(0, 1, 2, 2, 1, 0, 0, 1, 2, 2, 1, 0, 0, 1, 2, 2, 1, 0)'$$

to the array constructed above yields an $OA(18, 6, 3, 2)$ that is trend-free of order $(1, 1)$. ■

We now turn to the method of construction via the use of Galois fields, as discussed in Section 3.4. Then the following result, due to Coster and Cheng (1988; see also Wang, 1991) holds:

Theorem 7.3.3. *Let $m(\geq 2)$ be a prime or a prime power. Suppose that there exists an $r \times n$ matrix C, with elements from $GF(m)$, such that every $r \times g$ submatrix of C has rank g and every column of C has at least $s+1$ nonzero elements. Then there exists a symmetric $OA(m^r, n, m, g)$ that is trend-free of order $(1, s)$.*

Proof. Let $\rho_0, \rho_1, \ldots, \rho_{m-1}$ be the elements of $GF(m)$ with $\rho_0(= 0)$ denoting the identity element with respect to the operation of addition. Denote the rows of C by $\boldsymbol{c}_1', \ldots, \boldsymbol{c}_r'$. Construct an $m^r \times n$ array A such that for $0 \leq u_1, \ldots, u_r \leq m - 1$, the $(u_1 m^{r-1} + u_2 m^{r-2} + \cdots + u_r + 1)$th row of A is given by $\sum_{i=1}^{r} \rho_{u_i} \boldsymbol{c}_i'$. Then A is as described in the proof of Lemma 3.4.1. Therefore, noting that every $r \times g$ submatrix of C has rank g, it follows that A is an $OA(m^r, n, m, g)$.

By the above construction, the jth column of A is given by

$$\boldsymbol{a}_j = \boldsymbol{h}_{1j} \odot \boldsymbol{h}_{2j} \odot \cdots \odot \boldsymbol{h}_{rj}, \qquad 1 \leq j \leq n, \tag{7.3.14}$$

where

$$\boldsymbol{h}_{ij} = (\rho_0 c_{ij}, \rho_1 c_{ij}, \ldots, \rho_{m-1} c_{ij})', \qquad 1 \leq i \leq r, \tag{7.3.15}$$

and $(c_{1j}, c_{2j}, \ldots, c_{rj})'$ is the jth column of C. By (7.3.15), $\boldsymbol{h}_{ij}$ equals the null vector if $c_{ij} = 0$, while if $c_{ij} \neq 0$, then $\boldsymbol{h}_{ij}$ is 0-trend-free. Since each column of C has at least $s + 1$ nonzero elements, by repeated application of Lemma 7.3.1, it now follows from (7.3.14) that $\boldsymbol{a}_j$ is s-trend-free for every $j(1 \leq j \leq n)$. Thus the array A is trend-free of order $(1, s)$. ■

Because of the similarity between (7.3.10) and (7.3.14) (see also (7.3.15)), the method of construction in Theorem 7.3.3 is called the generalized foldover technique. In the spirit of Chatterjee and Mukerjee (1986b), we get the following corollary from Theorem 7.3.3:

Corollary 7.3.1. *Let $m(\geq 2)$ be a prime or a prime power and r, s be integers such that (i) $r \geq 3, 1 \leq s \leq r-2$, if $m = 2$, and (ii) $r \geq 2, 1 \leq s \leq r-1$,* *if $m \geq 3$. Define*

$$n_{rs} = \frac{m^r - 1}{m - 1} - \sum_{j=1}^{s} \binom{r}{j} (m - 1)^{j-1}.$$

Then for $2 \leq n \leq n_{rs}$, there exists a symmetric $OA(m^r, n, m, 2)$ that is trend-free of order $(1, s)$.

Proof. Define the set of $r \times 1$ vectors T as in the proof of Theorem 3.4.3. Recall that T has cardinality $(m^r - 1)/(m - 1)$ and that every two distinct vectors in T are linearly independent. As noted in Remark 3.4.3, the vectors in T represent the distinct points of the finite projective geometry $PG(r-1, m)$. For $1 \leq j \leq r$, it is easy to see that the number of vectors in T with exactly j nonzero elements equals $\binom{r}{j}(m - 1)^{j-1}$. Hence, with s as in (i) or (ii) above, there are n_{rs} vectors in T each having at least $s + 1$ nonzero elements. Since $n \leq n_{rs}$, the result now follows from Theorem 7.3.3 taking C as an $r \times n$ matrix with columns given by any n of these n_{rs} vectors. ■

Incidentally, the conditions (i) and (ii) in the statement of Corollary 7.3.1 are not restrictive. They only ensure that $n_{rs} \geq 2$, so that the result is meaningful.

Example 7.3.2. (John, 1990). In Corollary 7.3.1, take $m = r = 3$, $s = 2$. Then $n_{rs} = 4$ and the four vectors in T each having three nonzero elements are $(1, 1, 1)'$, $(1, 2, 1)'$, $(1, 1, 2)'$, $(1, 2, 2)'$. Let $n = 4$. Then with

$$C = \begin{bmatrix} 1 & 1 & 1 & 1 \\ 1 & 2 & 1 & 2 \\ 1 & 1 & 2 & 2 \end{bmatrix},$$

proceeding as in the proof of Theorem 7.3.3, one gets the array

$$\begin{bmatrix} 012 & 120 & 201 & 120 & 201 & 012 & 201 & 012 & 120 \\ 012 & 201 & 120 & 120 & 012 & 201 & 201 & 120 & 012 \\ 021 & 102 & 210 & 102 & 210 & 021 & 210 & 021 & 102 \\ 021 & 210 & 102 & 102 & 021 & 210 & 210 & 102 & 021 \end{bmatrix}'.$$

Here not only every two distinct columns of C are linearly independent but also every 3×3 submatrix of C has rank 3. Hence the above array is an $OA(27, 4, 3, 3)$ which is trend-free of order $(1, 2)$. In contrast, the $OA(27, 4, 3, 3)$ shown in Example 3.4.1 is not even trend-free of order $(1, 1)$. ■

In particular, for $m = 2$, the following stronger version of Theorem 7.3.3, due to Cheng and Jacroux (1988), holds:

Theorem 7.3.4. *Suppose that there exists an $r \times n$ matrix C, with elements from $GF(2)$, such that (i) every $r \times g$ submatrix of C has rank g and (ii) for $1 \le f' \le f$, the sum of every f' distinct columns of C has at least $s + 1$ elements equal to 1, where $f(\le g)$ is a fixed positive integer. Then there exists a symmetric $OA(2^r, n, 2, g)$ which is trend-free of order (f, s).*

Proof. For notational simplicity, we prove the result for $f = 2$. The treatment is similar for $f \ge 3$ while the case $f = 1$ is covered by Theorem 7.3.3. Denote the rows of C by $\boldsymbol{c}_1', \ldots, \boldsymbol{c}_r'$. As in the proof of Theorem 7.3.3, construct a $2^r \times n$ array A such that for $u_1, \ldots, u_r \in \{0, 1\}$, the $(u_1 2^{r-1} + u_2 2^{r-2} + \cdots + u_r + 1)$th row of A is given by $\sum_{i=1}^{r} u_i \boldsymbol{c}_i'$. Then by condition (i), A is an $OA(2^r, n, 2, g)$.

Denote the columns of A by $\boldsymbol{a}_1, \ldots, \boldsymbol{a}_n$, and let d_0 be the plan represented by A. Let $N = 2^r$. Comparing the notation used in (7.2.7) and (7.3.1), observe that

$$\chi_{\boldsymbol{a}_i}(u; j) = \chi_{d_0}{}^{i}(u; j), \tag{7.3.16}$$

for $1 \le i \le n$, $1 \le u \le N$ and $j = 0, 1$. Since condition (ii) holds with $f = 2$, as in the proof of Theorem 7.3.3, from our construction it follows that $\boldsymbol{a}_1$, $\boldsymbol{a}_2$ and $\boldsymbol{a}_1 + \boldsymbol{a}_2$ are all s-trend-free. Since $\boldsymbol{a}_1$ and $\boldsymbol{a}_2$ are s-trend-free, by (7.3.1) and (7.3.16),

$$\sum_{u=1}^{N} u^k \chi_{d_0}{}^{i}(u; 0) = \sum_{u=1}^{N} u^k \chi_{d_0}{}^{i}(u; 1), \qquad 1 \le k \le s, \quad i = 1, 2. \tag{7.3.17}$$

Also any element of $\boldsymbol{a}_1 + \boldsymbol{a}_2$ equals zero if the corresponding elements of $\boldsymbol{a}_1$ and $\boldsymbol{a}_2$ are equal, and 1 otherwise. Since $\boldsymbol{a}_1 + \boldsymbol{a}_2$ is s-trend-free, we get

$$\sum_{u=1}^{N} u^k \{\chi_{d_0}{}^{12}(u; 00) + \chi_{d_0}{}^{12}(u; 11)\} = \sum_{u=1}^{N} u^k \{\chi_{d_0}{}^{12}(u; 01) + \chi_{d_0}{}^{12}(u; 10)\},$$
$$1 \le k \le s. \tag{7.3.18}$$

Next observe that for $j = 0, 1$,

$$\chi_{d_0}{}^{1}(u; j) = \chi_{d_0}{}^{12}(u; j0) + \chi_{d_0}{}^{12}(u; j1),$$
$$\chi_{d_0}{}^{2}(u; j) = \chi_{d_0}{}^{12}(u; 0j) + \chi_{d_0}{}^{12}(u; 1j).$$

Hence for $1 \le k \le s$, denoting the 4×1 vector with elements

$$\sum_{u=1}^{N} u^k \chi_{d_0}{}^{12}(u; j_1 j_2) \; (j_1, j_2 = 0, 1),$$

arranged in the lexicographic order, by $\boldsymbol{\chi}^{12}(k)$, it follows from (7.3.17) and (7.3.18) that

$$\begin{bmatrix} 1 & 1 & -1 & -1 \\ 1 & -1 & 1 & -1 \\ 1 & -1 & -1 & 1 \end{bmatrix} \boldsymbol{\chi}^{12}(k) = \mathbf{0}.$$

This shows that the elements of $\boldsymbol{\chi}^{12}(k)$ are all equal. Since the elements of $\boldsymbol{\chi}^{12}(k)$ add up to $\sum_{u=1}^{N} u^k$, clearly each of them equals $\frac{1}{4}\sum_{u=1}^{N} u^k$. Similarly it can be shown that (7.2.7) holds for every i_1, i_2 and j_{i_1}, j_{i_2} $(1 \le i_1 < i_2 \le n;$ $j_{i_1}, j_{i_2} = 0, 1)$. Hence the array A is trend-free of order $(2, s)$. ■

Cheng and Jacroux (1988) considered several applications of Theorem 7.3.4. One of these is presented in the next example.

Example 7.3.3. Let $n(\ge 7)$ be odd and, for $1 \le j \le n-1$, $\boldsymbol{e}_j$ be the $(n-1) \times 1$ vector over $GF(2)$ with jth element 1 and all other elements zero. Let $C = [\boldsymbol{c}^{(1)}, \boldsymbol{c}^{(2)}, \ldots, \boldsymbol{c}^{(n)}]$ be an $(n-1) \times n$ matrix, with elements from $GF(2)$, such that

$$\boldsymbol{c}^{(i)} = \sum_{j=1}^{n-3} \boldsymbol{e}_j - \boldsymbol{e}_i \qquad (i = 1, 2, \ldots, n-3),$$

$$\boldsymbol{c}^{(i)} = \sum_{j=1}^{n-5} \boldsymbol{e}_j + \boldsymbol{e}_i \qquad (i = n-2, n-1),$$

$$\boldsymbol{c}^{(n)} = \sum_{j=1}^{n-1} \boldsymbol{e}_j.$$

Then every $(n-1) \times (n-1)$ submatrix of C has rank $n-1$. Furthermore every column of C has at least $n-4$ ones, and the sum of every two distinct columns of C has at least 2 ones. Hence following Theorem 7.3.4, from C one can construct an $OA(2^{n-1}, n, 2, n-1)$ that is trend-free of order $(1, n-5)$ and also of order $(2, 1)$. ■

Theorems 7.3.1 and 7.3.4 integrate some of the early results reported in Hill (1960) and Daniel and Wilcoxon (1966). For an illuminating discussion on the related developments, the reader should see Cheng (1990). We also refer to Coster and Cheng (1988), Bailey et al. (1992), Coster (1993a), and Jacroux (1994, 1996) for further extensions of the results presented in this section. The last four of these papers present results pertaining to asymmetric factorials as well. Along the line of Chapter 5, it is also of interest to study the issue of nonexistence of trend-free orthogonal arrays. However, this problem is yet to receive attention in the literature.

The results discussed thus far in this chapter relate to a fixed effects model with uncorrelated, homoscedastic observations, the effect of a time trend being represented by a polynomial. More complex situations were considered by Cheng and Steinberg (1991). They studied the problem of finding trend-robust run orders of two-level factorials when time effects are modeled via a first-order autoregressive error model or a time series model; see also Steinberg (1988) in this context. More results in this direction were reported in Saunders et al. (1995).

Before concluding this section, we briefly discuss another related issue. There can be experimental situations where it is expensive to change factor levels from one run to the next. For example, the levels of a factor may be the temperature settings of a furnace, and it may be expensive to change the temperature from one run to another. Such situations warrant the use of run orders that minimize the cost of level changes (between consecutive runs) which can be measured by the number of level changes when all level changes are equally expensive. Early work in this area is due to Draper and Stoneman (1968), Dickinson (1974), and Joiner and Campbell (1976). Cheng (1985) was the first to study the problem in a systematic manner in the context of 2^n factorials.

If the observations are also influenced by a time trend, then the twin objectives of trend elimination, as discussed so far in this chapter, and minimization of the number of level changes may not in general lead to the same run order; see Cheng (1985) for a discussion. However, Cheng (1985) also observed that there can be situations where the two criteria are in agreement. The following example illustrates an interesting situation of this kind:

Example 7.3.4. (Cheng, 1985). Consider the setup of a 2^n factorial where $n \geq 5$. Let $C = [\boldsymbol{c}^{(1)}, \boldsymbol{c}^{(2)}, \ldots, \boldsymbol{c}^{(n)}]$ be an $(n-1) \times n$ matrix, with elements from $GF(2)$, such that

$$\begin{aligned}
\boldsymbol{c}^{(i)} &= \boldsymbol{e}_{n-i-1} + \boldsymbol{e}_{n-i} \qquad (i = 1, 2, \ldots, n-3), \\
\boldsymbol{c}^{(n-2)} &= \boldsymbol{e}_1 + \boldsymbol{e}_2 + \boldsymbol{e}_3, \\
\boldsymbol{c}^{(n-1)} &= \boldsymbol{e}_1 + \boldsymbol{e}_2, \\
\boldsymbol{c}^{(n)} &= \boldsymbol{e}_2 + \boldsymbol{e}_3 + \boldsymbol{e}_{n-1},
\end{aligned} \tag{7.3.19}$$

where, as before, $\boldsymbol{e}_1, \ldots, \boldsymbol{e}_{n-1}$ are the unit vectors over $GF(2)$. Denote the rows of C by $\boldsymbol{c}_1', \ldots, \boldsymbol{c}_{n-1}'$, and as in the proof of Theorem 7.3.4, construct a $2^{n-1} \times n$ array A such that for $u_1, \ldots, u_{n-1} \in \{0, 1\}$, the $(u_1 2^{n-2} + u_2 2^{n-3} + \cdots + u_{n-1} + 1)$th row of A is given by $\sum_{i=1}^{n-1} u_i \boldsymbol{c}_i'$. For example, if $n = 5$, then

$$C = \begin{bmatrix} 0 & 0 & 1 & 1 & 0 \\ 0 & 1 & 1 & 1 & 1 \\ 1 & 1 & 1 & 0 & 1 \\ 1 & 0 & 0 & 0 & 1 \end{bmatrix}$$

and

$$A = \begin{bmatrix} 0110 & 0110 & 0110 & 0110 \\ 0011 & 1100 & 0011 & 1100 \\ 0011 & 1100 & 1100 & 0011 \\ 0000 & 1111 & 1111 & 0000 \\ 0110 & 1001 & 0110 & 1001 \end{bmatrix}'. \tag{7.3.20}$$

By (7.3.19), every $(n-1) \times (n-1)$ submatrix of C has rank $n-1$, and every column of C has at least two 1's. Hence, by Theorem 7.3.4, the array A constructed from C as above is an $OA(2^{n-1}, n, 2, n-1)$ which is trend-free of order $(1, 1)$. From Corollary 7.2.1, it follows that the plan d_0 given by A is a universally optimal Resolution $(1, n-2)$ plan in $\mathcal{D}_N$, where $N = 2^{n-1}$, in the presence of a linear trend.

Interestingly d_0 also minimizes the number of level changes within the class of all plans that are represented by an $OA(2^{n-1}, n, 2, n-1)$. To see this, observe that the rows of every $2^{n-1} \times (n-1)$ subarray of an $OA(2^{n-1}, n, 2, n-1)$ are all distinct. Therefore every two consecutive rows of an $OA(2^{n-1}, n, 2, n-1)$ differ in at least two positions so that in a plan, given by such an array, there are at least $2(2^{n-1} - 1)$ level changes between consecutive runs. The plan d_0 attains this lower bound, since under the present construction every two consecutive rows of A differ in exactly two positions (see (7.3.20)). ■

Considering symmetric m^n factorials, where m is a prime or a prime power, Coster and Cheng (1988) addressed in more detail the problem of finding fractional factorial plans that minimize the cost of level changes while remaining trend-free. Coster (1993b) presented tables of such plans for 2^n and 3^n factorials. Further results in this area were reported by Wang (1991, 1996) and El Mossadeq and Kobilinsky (1992).

7.4 BLOCKING

So far we have worked under the tacit assumption of homogeneity of the experimental units. Especially when the number of runs, N, is large, it may be difficult to ensure such homogeneity. A common practice in situations of this kind is to group the experimental units into classes, known as blocks, so that the units within each block are homogeneous, though there may be variation from one block to another. This calls for consideration of a linear model, via a suitable modification of the model (2.3.12) introduced in Chapter 2, so as to incorporate the block effects.

To be more specific, with reference to an $m_1 \times \cdots \times m_n$ factorial, let $\mathcal{D}(b, k)$ be the class of plans involving $N(= bk)$ experimental units grouped into $b(\geq 2)$ blocks each containing $k(\geq 2)$ units. As before, suppose that interest lies in the general mean and factorial effects involving f or less factors under the absence of factorial effects involving $t + 1$ or more factors $(1 \leq f \leq t \leq n - 1)$. At this stage we do not consider the presence of any time trend. Consider any plan $d \in \mathcal{D}(b, k)$. For $1 \leq u \leq N$, $1 \leq l \leq b$, let the indicator $z_d(u, l)$ assume the value 1 if the uth observation according to d comes from a unit in the lth block and the value zero otherwise. Let Z_d be an $N \times b$ matrix with (u, l)th element $z_d(u, l)$, and write $\boldsymbol{\theta}^* = (\theta_1{}^*, \ldots, \theta_b{}^*)'$, where $\theta_1{}^*, \ldots, \theta_b{}^*$ are parameters representing the effects of the b blocks. Then the model, associated with the plan d, may be taken as

$$\mathbb{E}(\boldsymbol{Y}) = X_{1d}\boldsymbol{\beta}^{(1)} + X_{2d}\boldsymbol{\beta}^{(2)} + Z_d\boldsymbol{\theta}^*, \quad \mathbb{D}(\boldsymbol{Y}) = \sigma^2 I_N, \tag{7.4.1}$$

where $\boldsymbol{Y}$ is the observational vector and X_{1d}, X_{2d}, $\boldsymbol{\beta}^{(1)}$, $\boldsymbol{\beta}^{(2)}$, σ^2, and I_N are exactly as in (2.3.12) or (7.2.2). In particular, $\boldsymbol{\beta}^{(1)}$ represents the parametric vector of interest.

Without loss of generality, it may be supposed that the block effects $\theta_1{}^*, \ldots, \theta_b{}^*$ are measured from their mean which, being common to all the observations, is absorbed in the general mean. Then $\theta_1{}^* + \cdots + \theta_b{}^* = 0$, which induces a reparametrization $\boldsymbol{\theta}^* = Q'\boldsymbol{\theta}$ in terms of a parametric vector $\boldsymbol{\theta} = (\theta_1, \ldots, \theta_{b-1})'$, where Q is a $(b - 1) \times b$ matrix of full row rank, satisfying

$$Q\mathbf{1}_b = \mathbf{0}. \tag{7.4.2}$$

The model (7.4.1) can now be expressed as

$$\mathbb{E}(\boldsymbol{Y}) = X_{1d}\boldsymbol{\beta}^{(1)} + X_{2d}\boldsymbol{\beta}^{(2)} + Z_d Q'\boldsymbol{\theta}, \qquad \mathbb{D}(\boldsymbol{Y}) = \sigma^2 I_N. \tag{7.4.3}$$

Let $\mathcal{I}_d{}^{\text{block}}$ be the information matrix for $\boldsymbol{\beta}^{(1)}$ under the plan d and the model (7.4.3). Depending on whether $f < t$ or $f = t$, the expressions for $\mathcal{I}_d{}^{\text{block}}$ are as in (7.2.4) or (7.2.5), respectively, with X_3 replaced by $Z_d Q'$. As in Theorem 2.3.1, $\boldsymbol{\beta}^{(1)}$ is estimable under d and the model (7.4.3) if and only if $\mathcal{I}_d{}^{\text{block}}$ is positive definite, in which case d will be called a Resolution (f, t) plan. Note that $\mathcal{I}_d - \mathcal{I}_d{}^{\text{block}}$ is nonnegative definite, where $\mathcal{I}_d$, given by (2.3.14), is the information matrix for $\boldsymbol{\beta}^{(1)}$ under d and the model (2.3.12). Therefore the same consideration as in Section 7.2 applies while studying the issue of optimality in the present setup.

Theorem 7.4.1. *Let $d_0 \in \mathcal{D}(b, k)$ be represented by an orthogonal array $OA(N, n, m_1 \times \cdots \times m_n, f + t)$.*

(a) *Then $\mathcal{I}_{d_0}{}^{\text{block}} = \mathcal{I}_{d_0}$ if and only if for every $i_1, \ldots, i_f$ and $j_{i_1}, \ldots, j_{i_f}$ $(1 \le i_1 < \cdots < i_f \le n; 0 \le j_{i_1} \le m_{i_1} - 1, \ldots, 0 \le j_{i_f} \le m_{i_f} - 1)$, the level combination $j_{i_1} \ldots j_{i_f}$ of the i_1th, $\ldots$, i_fth factors appears equally often in the b blocks under the plan d_0.*

(b) *If the condition stated in (a) above holds, then d_0 is a universally optimal Resolution (f, t) plan in $\mathcal{D}(b, k)$.*

Proof.

(a) As in the proof of Theorem 7.2.1, $\mathcal{I}_{d_0}{}^{\text{block}} = \mathcal{I}_{d_0}$ if and only if

$$X_{1d_0}{}' Z_{d_0} Q' = O. \tag{7.4.4}$$

Hence we need to show the equivalence of (7.4.4) and the stated condition. For notational simplicity, this will be done for $f = 2$; the treatment is similar for general f.

Let (7.4.4) hold. Then by (7.2.3), $P^{(1)} X_{d_0}{}' Z_{d_0} Q' = O$, and since $f = 2$, the same argument as in Theorem 7.2.1 yields

$$[I_{m_1} \otimes I_{m_2} \otimes \mathbf{1}'_{m_3} \otimes \cdots \otimes \mathbf{1}'_{m_n}] X_{d_0}{}' Z_{d_0} Q' = O.$$

Since the $(b - 1) \times b$ matrix Q, of full row rank, satisfies (7.4.2), the above implies that the matrix

$$U = [I_{m_1} \otimes I_{m_2} \otimes \mathbf{1}'_{m_3} \otimes \cdots \otimes \mathbf{1}'_{m_n}] X_{d_0}{}' Z_{d_0}$$

must have all elements identical in every row. But by the definitions of X_{d_0} and Z_{d_0}, the elements in any row of the $m_1 m_2 \times b$ matrix U represent the frequencies with which the corresponding level combination of the first two factors appears in the b blocks under the plan d_0. Hence under d_0, every level combination of the first two factors occurs equally often in the b blocks. Similarly it can be seen that (7.4.4) leads to the same conclusion for the level combinations of any other pair of factors. Thus (7.4.4) implies the stated condition. The converse can be proved by reversing the above steps.

(b) Follows as in the proof of Theorem 7.2.1 noting that $\mathcal{I}_d - \mathcal{I}_d^{\text{block}}$ is nonnegative definite for every $d \in \mathcal{D}(b, k)$. ■

Example 7.4.1. With reference to a 2^4 factorial, let $N = 8$, $b = 4$, $k = 2$, and consider the plan $d_0 \in \mathcal{D}(4, 2)$ as shown below:

	Block 1	Block 2	Block 3	Block 4
$d_0 \equiv$	0000	0011	0101	0110
	1111	1100	1010	1001

Then d_0 is represented by an $OA(8, 4, 2, 3)$. Furthermore every level of every factor appears exactly once in each block. Hence from Theorem 7.4.1 it follows that d_0 is a universally optimal Resolution $(1, 2)$ plan in $\mathcal{D}(4, 2)$. ■

In particular, with $N = bk$, suppose that an $OA(N, n+1, m_1 \times \cdots \times m_n \times b, f+t)$, say A, is available. One may identify each row of the subarray given by the first n columns with a treatment combination of an $m_1 \times \cdots \times m_n$ factorial and each symbol in the last column with a block. If a typical row of A is $j_1 \ldots j_n l$, then this assigns the treatment combination $j_1 \ldots j_n$ to the block l. Let $d_0 \in \mathcal{D}(b, k)$ be the resulting plan for an $m_1 \times \cdots \times m_n$ factorial. Clearly, d_0 is represented by an $OA(N, n, m_1 \times \cdots \times m_n, f+t)$ and satisfies the condition in part (a) of Theorem 7.4.1. Hence d_0 is a universally optimal Resolution (f, t) plan in $\mathcal{D}(b, k)$.

Example 7.4.2. Let A be the $OA(8, 5, 4 \times 2^4, 2)$ shown in Example 2.6.1(a). Identify each row of the subarray given by the first four columns of A with a treatment combination of a 4×2^3 factorial and each symbol in the last column of A with a block. This gives a fractional factorial plan d_0 arranged in two blocks

$$\{0000, 1110, 2101, 3011\} \qquad \{0111, 1001, 2010, 3100\}$$

each of four units. By Theorem 7.4.1, d_0 is a universally optimal Resolution $(1, 1)$ plan in $\mathcal{D}(2, 4)$. ■

The problem of blocking has been studied in some more detail in the special case of symmetric m^n factorials where m is a prime or a prime power.

We will return to this point in Chapter 8. The interested reader should also see Bailey (1977), Morgan and Uddin (1996), and the references therein for further related results. Several authors have explored the issue of optimality in settings that incorporate both block effects (possibly with more complex block structures than discussed here) and time trend. We refer to Coster and Cheng (1988), Jacroux and Saharay (1990), and Coster (1993a) in this connection.

EXERCISES

7.1. Examine how the "if" part of Theorem 7.2.1(a) follows by reversing the steps in the proof of the "only if" part.

7.2. Let d_0 be represented by an orthogonal array $OA(N, n, m_1 \times \cdots \times m_n, 2f)$. If d_0 is s-trend-free, then show that $N \geq \alpha_f + s$, where α_f is defined as in (2.3.11). [*Hint*: Use (7.2.9).]

7.3. Verify parts (a) and (b) of Lemma 7.3.1.

7.4. Let $m(\geq 3)$ be any prime or prime power. Examine if Example 7.3.2 can be generalized to construct an $OA(m^3, 4, m, 3)$ that is trend-free of order (1, 2).

7.5. For any integer $r(\geq 5)$, show that there exists an $OA(2^r, 2^{r-1} - r, 2, 3)$ that is trend-free of order (1, 2). [*Hint*: Consider $r \times 1$ vectors over $GF(2)$, having an odd number (≥ 3) of 1's, as columns of the matrix C in Theorem 7.3.3.]

7.6. Assume that a difference matrix $D(\lambda m, n; m)$ exists. For an m^n factorial, indicate the construction of a universally optimal Resolution (1, 1) plan in $\mathcal{D}(\lambda m, m)$.

CHAPTER 8

Some Further Developments

8.1 INTRODUCTION

In this book we have so far been concerned with optimal fractional factorial plans under specified linear models. The present chapter differs in spirit from the previous ones in the sense of allowing a greater flexibility in the model. Our objective here is to present a concise introduction to the basics of certain newly emerging areas such as minimum aberration, search, and supersaturated designs. In the process we will come across new kinds of optimality criteria. We refer to Hamada and Wu (1999) for further discussion on some of the topics covered in this chapter.

Section 8.2 begins with a discussion on the classical regular fractions and discusses minimum aberration designs. Search and supersaturated designs are then briefly discussed in Sections 8.3 and 8.4, respectively.

8.2 REGULAR FRACTIONS AND MINIMUM ABERRATION DESIGNS

Consider the setup of a symmetric m^n factorial where $m(\geq 2)$ is a prime or a prime power. Following Bose (1947), it is then possible to develop an approach to fractional factorial plans providing a significant insight into the underlying issues. This utilizes the properties of a finite field and calls for the use of a notational system different from the one employed in the earlier chapters.

Let $\rho_0, \rho_1, \ldots, \rho_{m-1}$ be the elements of $GF(m)$, where $\rho_0(= 0)$ and $\rho_1(= 1)$ are the identity elements with respect to the operations of addition and multiplication respectively. A typical treatment combination $j_1 \ldots j_n$ of an m^n factorial can be identified with the vector $(\rho_{j_1}, \ldots, \rho_{j_n})'$ $(0 \leq j_1, \ldots, j_n \leq m-1)$. The m^n treatment combinations are thus represented by the m^n vectors of the form $z = (z_1, \ldots, z_n)'$, where $z_i \in GF(m)$

for each i. In other words, the treatment combinations are identified with the points of the n-dimensional finite Euclidean geometry $EG(n, m)$, based on $GF(m)$ (e.g., see Raghavarao, 1971, pp. 357–359). The effect of a treatment combination represented by z will be denoted by $\tau(z)$.

For symmetric prime or prime powered factorials, there is a convenient way of representing treatment contrasts belonging to factorial effects. Some preliminaries are needed in this context. Let $\boldsymbol{a} = (a_1, \ldots, a_n)'$ be any fixed nonnull vector over $GF(m)$. Then it is easy to see that each of the sets

$$V_j(\boldsymbol{a}) = \{z = (z_1, \ldots, z_n)' : \boldsymbol{a}'z = \rho_j\}, \qquad 0 \le j \le m-1, \tag{8.2.1}$$

has cardinality m^{n-1}, and that these sets provide a disjoint partition of the class of all treatment combinations. The sets in (8.2.1) are collectively called a parallel pencil of $(n-1)$-flats of $EG(n, m)$, and hence $\boldsymbol{a}$ itself is said to represent a pencil. Let

$$S_j(\boldsymbol{a}) = \sum_{z \in V_j(\boldsymbol{a})} \tau(z), \qquad 0 \le j \le m-1. \tag{8.2.2}$$

Then a treatment contrast L is said to belong to the pencil $\boldsymbol{a}$ if it is of the form

$$L = \sum_{j=0}^{m-1} l_j S_j(\boldsymbol{a}), \tag{8.2.3}$$

where $l_0, \ldots, l_{m-1}$ are real numbers, not all zeros, satisfying $\sum_{j=0}^{m-1} l_j = 0$. Clearly, there are $m-1$ linearly independent treatment contrasts belonging to the pencil $\boldsymbol{a}$.

It is easy to see from (8.2.1) that two pencils, say $\boldsymbol{a}$ and $\boldsymbol{a}^*$, satisfying $\boldsymbol{a}^* = \lambda\boldsymbol{a}$ for some $\lambda(\neq 0) \in GF(m)$, induce the same disjoint partition of the class of all treatment combinations. As such, hereafter, pencils with proportional entries will be considered as identical. Thus there are $(m^n - 1)/(m-1)$ distinct pencils, no two of which have proportional entries. For any two distinct pencils, $\boldsymbol{a}$, $\boldsymbol{b}$ and $0 \le j, j' \le m-1$, it follows from (8.2.1) that the set $V_j(\boldsymbol{a}) \cap V_{j'}(\boldsymbol{b})$ has cardinality m^{n-2}. Hence from (8.2.2) and (8.2.3), the following lemma can be proved easily.

Lemma 8.2.1. *Treatment contrasts belonging to distinct pencils are mutually orthogonal.* ■

We are now in a position to connect pencils with factorial effects. A pencil $\boldsymbol{a} = (a_1, \ldots, a_n)'$ is said to belong to a factorial effect $F_{i_1} \ldots F_{i_g}$ $(1 \le i_1 <$

$\cdots < i_g \le n;\ 1 \le g \le n$) provided that

$$\begin{aligned} a_i &\neq 0 \quad \text{if } i = i_1, \ldots, i_g, \\ &= 0 \quad \text{otherwise.} \end{aligned} \tag{8.2.4}$$

Comparing (8.2.1)–(8.2.4) with the contents of Section 1.2 (see the discussion following (1.2.1)), it is not hard to see that if a pencil $\boldsymbol{a}$ belongs to a factorial effect $F_{i_1} \ldots F_{i_g}$, then every treatment contrast belonging to $\boldsymbol{a}$ also belongs to $F_{i_1} \ldots F_{i_g}$. Now, pencils with proportional entries are identical, and hence by (8.2.4), there are $(m-1)^{g-1}$ distinct pencils belonging to $F_{i_1} \ldots F_{i_g}$. Each of them carries $m-1$ linearly independent treatment contrasts. Thus, in view of Lemma 8.2.1, these pencils provide a representation for the treatment contrasts belonging to $F_{i_1} \ldots F_{i_g}$ in terms of $(m-1)^{g-1}$ mutually orthogonal sets of contrasts with $m-1$ linearly independent contrasts in each set. This accounts for a complete collection of $(m-1)^g$ linearly independent treatment contrasts belonging to the factorial effect $F_{i_1} \ldots F_{i_g}$.

For example, let $m = 3$ and $n = 2$. Then the pencils $(1, 0)'$ and $(0, 1)'$ belong to main effects F_1 and F_2, respectively, while interaction $F_1 F_2$ is represented by the two distinct pencils $(1, 1)'$ and $(1, 2)'$, each carrying two linearly independent treatment contrasts, the contrasts belonging to distinct pencils being mutually orthogonal.

A *regular* fraction of an m^n factorial, where $m (\geq 2)$ is a prime or a prime power, is specified by any $k (1 \leq k < n)$ linearly independent pencils, say $\boldsymbol{a}^{(1)}, \ldots, \boldsymbol{a}^{(k)}$, and consists of treatment combinations z satisfying

$$Az = \boldsymbol{c},$$

where A is a $k \times n$ matrix with rows $\boldsymbol{a}^{(i)\prime}$, $1 \leq i \leq k$, and $\boldsymbol{c}$ is a fixed $k \times 1$ vector over $GF(m)$. In what follows, the specific choice of $\boldsymbol{c}$ is inconsequential, and hence for notational simplicity we assume, without loss of generality, that $\boldsymbol{c} = \boldsymbol{0}$, the null vector over $GF(m)$. Then a regular fractional factorial plan is given by, say,

$$d(A) = \{z : Az = \boldsymbol{0}\}. \tag{8.2.5}$$

Since the rows of the $k \times n$ matrix A are given by linearly independent pencils, it is clear that $d(A)$ contains $N = m^{n-k}$ distinct treatment combinations. By (8.2.5), these m^{n-k} treatment combinations, or equivalently points of $EG(n, m)$, form a subgroup of $EG(n, m)$, the group operation being componentwise addition.

We now proceed to explore the properties of $d(A)$, assuming that the underlying linear model is given by (2.3.3) (with d there replaced by $d(A)$)

where the only cognizable effects are those due to treatment combinations. With reference to $d(A)$, a pencil $\boldsymbol{a}$ is called a defining pencil if

$$\boldsymbol{a} \in \mathcal{M}(A'), \tag{8.2.6}$$

where $\mathcal{M}(\cdot)$ denotes the column space of a matrix. Since pencils with proportional entries are identical, and A has full row rank, there are $(m^k-1)/(m-1)$ distinct defining pencils. For every defining pencil $\boldsymbol{a}$, by (8.2.5) and (8.2.6), $\boldsymbol{a}'z = 0$ whenever $z \in d(A)$ which, in view of (8.2.1), yields $d(A) \subset V_0(\boldsymbol{a})$. Therefore, from (8.2.2) and (8.2.3), the following result is evident:

Lemma 8.2.2. *No treatment contrast belonging to any defining pencil is estimable in $d(A)$.* ■

Lemmas 8.2.3–8.2.5 below help in understanding the status of $d(A)$ relative to pencils other than the defining ones. We need to introduce the important notion of alias sets for presenting these lemmas. Let $\mathcal{B}(\equiv \mathcal{B}(A))$ be the set of pencils which are not defining pencils. There are

$$\frac{m^n-1}{m-1} - \frac{m^k-1}{m-1} = \frac{m^k(m^{n-k}-1)}{m-1}$$

distinct pencils in $\mathcal{B}$. Any two members of $\mathcal{B}$, say $\boldsymbol{b}$ and $\boldsymbol{b}^*$, are said to be aliases of each other if

$$\boldsymbol{b} - \boldsymbol{b}^* \in \mathcal{M}(A'). \tag{8.2.7}$$

Since $A = \left(\boldsymbol{a}^{(1)}, \ldots, \boldsymbol{a}^{(k)}\right)'$ has full row rank, it is easily seen that any $\boldsymbol{b}(\in \mathcal{B})$ has m^k distinct aliases, namely

$$\boldsymbol{b}(\lambda_1, \ldots, \lambda_k) = \boldsymbol{b} + \sum_{i=1}^{k} \lambda_i \boldsymbol{a}^{(i)} \qquad (\lambda_1, \ldots, \lambda_k \in GF(m)), \tag{8.2.8}$$

including itself. In fact, by (8.2.7), the relationship of being aliased is an equivalence relation that partitions $\mathcal{B}$ into $(m^{n-k}-1)/(m-1)$ equivalence classes each of cardinality m^k. Any such equivalence class is called an alias set with reference to the fraction $d(A)$. Thus the pencils $\boldsymbol{b}(\lambda_1, \ldots, \lambda_k)$, defined in (8.2.8), constitute the alias set containing $\boldsymbol{b}$.

Lemma 8.2.3. *Let $\boldsymbol{b}$, $\boldsymbol{b}^*(\in \mathcal{B})$ be aliases of each other and*

$$L = \sum_{j=0}^{m-1} l_j S_j(\boldsymbol{b}) \quad \text{and} \quad L^* = \sum_{j=0}^{m-1} l_j S_j(\boldsymbol{b}^*) \tag{8.2.9}$$

be treatment contrasts belonging to $\boldsymbol{b}$ *and* $\boldsymbol{b}^*$ *respectively. Then the parts of* L *and* L^**, that involve only the treatment combinations included in* $d(A)$*, are identical.*

Proof. By (8.2.1), (8.2.2), (8.2.5), and (8.2.9), the part of L that involves only the treatment combinations included in $d(A)$ is given by, say,

$$L(A) = \sum_{j=0}^{m-1} l_j S_j(\boldsymbol{b}, A), \tag{8.2.10}$$

where

$$S_j(\boldsymbol{b}, A) = \sum_{z \in V_j(\boldsymbol{b}, A)} \tau(z), \tag{8.2.11}$$

and

$$V_j(\boldsymbol{b}, A) = V_j(\boldsymbol{b}) \cap d(A) = \{z : \boldsymbol{b}'z = \rho_j,\ Az = \boldsymbol{0}\}. \tag{8.2.12}$$

Similarly the part of L^* that involves only the treatment combinations included in $d(A)$ equals $L^*(A)$, which is obtained upon replacement of $\boldsymbol{b}$ by $\boldsymbol{b}^*$ everywhere in (8.2.10)–(8.2.12). Since $\boldsymbol{b}$ and $\boldsymbol{b}^*$ are aliases of each other, by (8.2.7) and (8.2.12), $V_j(\boldsymbol{b}, A) = V_j(\boldsymbol{b}^*, A)$ for every j, whence the result follows. ■

Remark 8.2.1. Contrasts with matching coefficients, such as L and L^* of (8.2.9), will be called corresponding contrasts. Lemma 8.2.3 shows that corresponding contrasts belonging to pencils that are members of the same alias set cannot be distinguished on the basis of the fraction $d(A)$. In this sense pencils belonging to the same alias set are said to be confounded with one another. Interestingly, since $\boldsymbol{b} \notin \mathcal{M}(A')$ in the setup of Lemma 8.2.3, the set $V_j(\boldsymbol{b}, A)$ considered in (8.2.12) has cardinality m^{n-k-1} for each j, so by (8.2.10), (8.2.11), $L(A)$ itself is a contrast involving the treatment combinations in $d(A)$. ■

Lemma 8.2.4. *Let* $\mathcal{B}^*$ *be an alias set and consider corresponding treatment contrasts* $\sum_{j=0}^{m-1} l_j S_j(\boldsymbol{b}^*)$ *for* $\boldsymbol{b}^* \in \mathcal{B}^*$*. Then*

$$\sum{}^{*} \sum_{j=0}^{m-1} l_j S_j(\boldsymbol{b}^*) = m^k \sum_{j=0}^{m-1} l_j S_j(\boldsymbol{b}, A), \tag{8.2.13}$$

where $\sum^*$ *denotes sum over* $\boldsymbol{b}^* \in \mathcal{B}^*$ *and* $\boldsymbol{b}$ *is any particular member of* $\mathcal{B}^*$*.*

Proof. Let Z be the set of all the m^n treatment combinations. For $0 \leq j \leq m-1$, $\boldsymbol{b}^* \in \mathcal{B}^*$ and $z \in Z$, define the indicator $\phi_j(\boldsymbol{b}^*, z)$ that assumes the value 1 if $z \in V_j(\boldsymbol{b}^*)$ and the value 0 otherwise; namely by (8.2.1),

$$\phi_j(\boldsymbol{b}^*, z) = \begin{cases} 1 & \text{if } (\boldsymbol{b}^*)'z = \rho_j, \\ 0 & \text{otherwise.} \end{cases} \tag{8.2.14}$$

From (8.2.2), (8.2.3), and (8.2.14),

$$\sum\nolimits^{*} \sum_{j=0}^{m-1} l_j S_j(\boldsymbol{b}^*) = \sum_{j=0}^{m-1} l_j \sum_{z \in Z} \left\{ \sum\nolimits^{*} \phi_j(\boldsymbol{b}^*, z) \right\} \tau(z). \tag{8.2.15}$$

But $\boldsymbol{b} \in \mathcal{B}^*$ and hence the pencils given by (8.2.8) constitute the set $\mathcal{B}^*$. Therefore, by (8.2.14), for any fixed j and z the quantity $\sum^* \phi_j(\boldsymbol{b}^*, z)$ equals the cardinality of the set

$$\{\boldsymbol{\lambda} = (\lambda_1, \ldots, \lambda_k)' : \boldsymbol{\lambda}' A z = \rho_j - \boldsymbol{b}'z \text{ and } \lambda_i \in GF(m) \text{ for each } i\}.$$

Consequently, by (8.2.5) and (8.2.12),

$$\sum\nolimits^{*} \phi_j(\boldsymbol{b}^*, z) = \begin{cases} m^k & \text{if } z \in V_j(\boldsymbol{b}, A), \\ 0 & \text{if } z \in d(A) - V_j(\boldsymbol{b}, A), \\ m^{k-1} & \text{if } z \notin d(A). \end{cases}$$

If one substitutes the above in (8.2.15), recalls (8.2.11), and uses the fact that $\sum_{j=0}^{m-1} l_j = 0$, then the result follows. ∎

From (8.2.10) and Remark 8.2.1, it is clear that the right-hand side of (8.2.13) is a contrast involving only the treatment combinations included in $d(A)$. Therefore the right-hand side and hence the left-hand side of (8.2.13) will be estimable in $d(A)$. In other words, while pencils belonging to the same alias set are confounded with one another (*vide* Remark 8.2.1), the sum of corresponding contrasts belonging to such pencils is estimable in $d(A)$. Thus any treatment contrast belonging to a pencil $\boldsymbol{b} (\in \mathcal{B})$ is estimable in $d(A)$ if and only if corresponding contrasts belonging to all other pencils that are aliased with $\boldsymbol{b}$ are ignorable.

We will say that a pencil is estimable in $d(A)$ if so is every treatment contrast belonging to it. Similarly, if every treatment contrast belonging to a pencil is ignorable, then the pencil itself will be called ignorable. Then from Lemmas 8.2.3, 8.2.4, and the discussion in the last paragraph, the following is evident.

Lemma 8.2.5. *A pencil* $\boldsymbol{b} \in \mathcal{B}$ *is estimable in* $d(A)$ *if and only if all other pencils that are aliased with* $\boldsymbol{b}$ *are ignorable.* ■

Example 8.2.1. With reference to a 3^5 factorial, consider a regular fraction $d(A)$ which is given by (8.2.5) with

$$A = \begin{pmatrix} 1 & 1 & 0 & 2 & 0 \\ 1 & 2 & 1 & 0 & 2 \end{pmatrix},$$

and consists of $N = 27$ distinct treatment combinations. Here $m = 3, n = 5$, $k = 2$, and by (8.2.6), the distinct defining pencils are

$$(1, 1, 0, 2, 0)', \ (1, 2, 1, 0, 2)', \ (1, 0, 2, 1, 1)' \text{ and } (0, 1, 1, 1, 2)'.$$

No main effect pencil (i.e., one that corresponds to a main effect and has exactly one nonzero entry) is a defining pencil. By (8.2.8), the alias set containing the main effect pencil $(1, 0, 0, 0, 0)'$ is given by

$$\begin{array}{lll} \{(1, 0, 0, 0, 0)', & (1, 1, 2, 0, 1)', & (0, 1, 2, 0, 1)', \\ (1, 2, 0, 1, 0)', & (0, 0, 1, 2, 2)', & (1, 2, 2, 2, 1)', \\ (0, 1, 0, 2, 0)', & (1, 1, 1, 1, 2)', & (1, 0, 1, 2, 2)'\}. \end{array}$$

Similarly one can find the alias sets containing all other main effect pencils and check that no two distinct main effect pencils have been aliased with each other. Hence, if all factorial effects involving two or more factors (i.e., all interactions) are absent, then the pencils belonging to these effects are ignorable, and therefore, by Lemma 8.2.5, all main effect pencils are estimable in the regular fraction $d(A)$ under consideration. ■

Note that in Example 8.2.1 there are three or more nonzero entries in each defining pencil. Hence the conclusion in this example regarding the estimability of all main effect pencils under the absence of all interactions is, in fact, a consequence of the following general result:

Theorem 8.2.1. *In a regular fraction, all contrasts belonging to factorial effects involving* f *or less factors are estimable under the absence of all factorial effects involving* $t + 1$ *or more factors* $(1 \le f \le t \le n - 1)$ *if and only if each defining pencil has at least* $f + t + 1$ *nonzero entries.*

Proof. Let $\boldsymbol{b}$ be any pencil involving f or less nonzero entries. By (8.2.4) and Lemmas 8.2.2 and 8.2.5, $\boldsymbol{b}$ is estimable in a regular fraction under the absence of all factorial effects involving $t + 1$ or more factors if and only if

(i) $\boldsymbol{b}$ is not a defining pencil, and

(ii) each other pencil aliased with $\boldsymbol{b}$ has at least $t + 1$ nonzero entries.

By (8.2.8), the above conditions hold for every pencil $\boldsymbol{b}$ having f or less nonzero entries if and only if each defining pencil has at least $f + t + 1$ nonzero entries. Hence the result follows. ■

In view of Theorem 8.2.1, the minimum number of nonzero entries in a defining pencil plays an important role in determining the behavior of a regular fraction. Following Box and Hunter (1961a,b), this minimum number is called the resolution of such a fraction. In Remark 8.2.2 below, we will indicate the link between this definition of resolution and what was introduced in Section 2.4 and followed in the previous chapters. Observe that each defining pencil in Example 8.2.1 has three or four nonzero entries. Hence the plan considered there has resolution three according to the present definition. In view of Theorem 8.2.1, a regular fraction of resolution one or two cannot ensure the estimability of all main effect contrasts even under the absence of all interactions. As such, hereafter, only regular fractions of resolution three or more will be considered. Then the following corollary is evident from Theorem 8.2.1:

Corollary 8.2.1. *A regular fraction of resolution $R(\geq 3)$ keeps all contrasts belonging to factorial effects involving f or less factors estimable under the absence of all factorial effects involving $R - f$ or more factors, whenever the integer f satisfies $1 \leq f \leq \frac{1}{2}(R - 1)$.* ■

The next result connects regular fractions with orthogonal arrays.

Theorem 8.2.2. *The treatment combinations included in a regular fraction of resolution R, when written as rows, represent a symmetric orthogonal array of strength $R - 1$.*

Proof. Consider a regular fraction $d(A)$ as given by (8.2.5). Since the $k \times n$ matrix A has full row rank, there exists an $(n-k) \times n$ matrix C, with elements from $GF(m)$ and also of full row rank, such that the row spaces of A and C are orthocomplements of each other, namely,

$$CA' = O. \tag{8.2.16}$$

By (8.2.5), then the m^{n-k} distinct vectors in the row space of C represent the treatment combinations in $d(A)$, written as rows. Furthermore, if $d(A)$ has resolution R, then every $(n - k) \times (R - 1)$ submatrix of C has rank $R - 1$; otherwise, by (8.2.16), there exists a nonnull vector in $\mathcal{M}(A')$, namely a defining pencil (see (8.2.6)), with less than R nonzero entries. This setting is precisely the same as that in Lemma 3.4.1, and the rest of the proof follows using the same arguments as there. ■

Remark 8.2.2. From Theorems 2.6.1 and 8.2.2, it is clear that a regular fraction of resolution $R(\geq 3)$ represents a universally optimal Resolution (f, t) plan, in the sense of Section 2.4 and within the class of all plans having the same number of runs, for every choice of integers f, t such that $f + t = R - 1$ and $1 \leq f \leq t \leq n - 1$. The reader may refer back to Remark 2.6.2 as well in this connection. ■

Remark 8.2.3. Theorem 8.2.2 shows that the necessary conditions for the existence of symmetric orthogonal arrays, notably Rao's bound, are also applicable to regular fractions. Further details on the issue of existence of regular fractions in the case of two-level factorials are available in Draper and Lin (1990). Interestingly, if $R \geq 3$ in Theorem 8.2.2, then no two distinct columns of the matrix C, of full row rank, are proportional, so the n columns of C can be interpreted as distinct points of the finite projective geometry $PG(n - k - 1, m)$ (see Remark 3.4.3). Because of the duality between the matrices C and A in the sense of having row spaces that are orthocomplements of each other, it follows that a regular fraction of resolution three or more is equivalent to a set of n distinct points of a finite projective geometry such that the matrix, given by these points as columns, has full row rank. Since $PG(n - k - 1, m)$ contains $(m^{n-k} - 1)/(m - 1)$ distinct points, given m, n, and k, then a regular fraction of resolution three or more exists if and only if

$$n \leq (m^{n-k} - 1)/(m - 1). \tag{8.2.17}$$

■

Corollary 8.2.1 suggests that given m, n, and k, one should look for a regular fraction having resolution as high as possible. However, there may be more than one regular fraction with the maximum possible resolution. Although they all share the same universal optimality property as indicated in Remark 8.2.2 under a specified model, one may discriminate among them on the basis of their defining pencils and aliasing patterns when interest lies in studying the consequences of possible model variation. To illustrate this point, we revisit Example 8.2.1. There $m = 3$, $n = 5$, $k = 2$, and no regular fraction can have resolution greater than three. This follows noting that if a fraction of resolution four is available, then, by Theorem 8.2.2, one gets a symmetric $OA(27, 5, 3, 3)$ which is nonexistent by Remark 3.4.1. The fraction discussed in Example 8.2.1 is, indeed, of the maximum possible resolution, namely three. However, in the same setup the fraction $d(\tilde{A})$, where

$$\tilde{A} = \begin{pmatrix} 1 & 1 & 0 & 2 & 0 \\ 1 & 0 & 1 & 0 & 2 \end{pmatrix},$$

has defining pencils

$$(1, 1, 0, 2, 0)', \quad (1, 0, 1, 0, 2)', \quad (1, 2, 2, 1, 1)', \quad (0, 1, 2, 2, 1)',$$

and is also of resolution three. Observe that $d(A)$ has only one defining pencil with exactly three nonzero entries, while there are two such pencils in $d(\tilde{A})$. Hence, by explicitly writing the alias sets or directly from (8.2.8), it can be seen that in $d(A)$ there are three distinct two-factor interaction pencils (i.e., pencils with exactly two nonzero entries) that are aliased with the main effect pencils, while in $d(\tilde{A})$ the number of such two-factor interaction pencils equals six. Therefore, if one is not fully confident about the absence of all two-factor interactions, then $d(A)$ seems to be preferable to $d(\tilde{A})$ although both of them have the highest possible resolution. More generally, if $W_3(\geq 0)$ be the number of pencils with exactly three nonzero entries in any regular fraction of resolution three or more, then from (8.2.8) it may be verified that $3W_3$ distinct two-factor interaction pencils get aliased in such a fraction with the main effect pencils. Thus, given any two resolution three fractions, the one with a smaller value of W_3 will be preferred under possible uncertainty about the model.

The considerations mentioned in the last paragraph lead to the criterion of minimum aberration introduced by Fries and Hunter (1980) for regular fractions. Under possible model uncertainty, this criterion is particularly meaningful when all pencils with the same number of nonzero entries are considered equally important, and as usual, the lower-order factorial effects are supposed to be more important than the higher-order ones. With reference to a regular fraction $d(A)$, as defined in (8.2.5), let $W_i(A)$ be the number of defining pencils with exactly i nonzero entries $(1 \leq i \leq n)$. Then the sequence

$$\{W_1(A), W_2(A), W_3(A), \ldots, W_n(A)\}$$

is called the word length pattern of $d(A)$ following a coding theoretic terminology. Given m, n, and k, consider two regular fractions $d(A)$ and $d(\tilde{A})$, and suppose that u is the smallest integer such that $W_u(A) \neq W_u(\tilde{A})$. Then $d(A)$ is said to have less aberration than $d(\tilde{A})$ if $W_u(A) < W_u(\tilde{A})$. A regular fraction $d(A)$ is said to represent a minimum aberration (MA) design if there is no other regular fraction having less aberration than $d(A)$. Obviously, an MA design has the highest possible resolution as well in a given context.

In the setup of Example 8.2.1, the word length patterns of $d(A)$ and $d(\tilde{A})$ are $\{0, 0, 1, 3, 0\}$ and $\{0, 0, 2, 1, 1\}$, respectively, so $d(A)$ has less aberration than $d(\tilde{A})$. In fact Chen, Sun, and Wu (1993) noted that $d(A)$ is the MA design here.

Franklin (1984) gave some tables of MA designs. Theoretical characterizations for such designs were obtained by Chen and Wu (1991) for $k = 3, 4$, and by Chen (1992) for $m = 2, k = 5$. Using an efficient computational algorithm, Chen, Sun, and Wu (1993) compiled a catalog of regular fractions that are good under the criterion of aberration. This catalog incorporates, in particular, MA designs for $m = 2$ and 3 over a practically useful range. Laycock and Rowley (1995) suggested a method of generating all regular fractions for given m, n, and k. Their results can help in finding MA designs.

The combinatorics underlying MA designs is deep and much new ground has recently been broken via consideration of complementary sets. From Remark 8.2.3, recall that a regular fraction of resolution three or more is equivalent to a set of n distinct points of the finite projective geometry $PG(n - k - 1, m)$. When n is close to the upper bound in (8.2.17), the complementary set in $PG(n-k-1, m)$ has a small cardinality. In this and several other situations, the theoretical study of MA designs can be greatly facilitated by expressing the word length pattern of a regular fraction in terms of the complementary set. The reader is referred to Chen and Hedayat (1996), Tang and Wu (1996), and Suen et al. (1997) for these interesting developments.

Wu and Zhang (1993) extended the idea of MA designs to the case of $2^{n_1} \times 4^{n_2}$ factorials constructed by the method of grouping discussed in Section 4.6. This work was followed up by Mukerjee and Wu (1997a).

In an attempt to provide more explicit statistical justification for MA designs, Cheng, Steinberg, and Sun (1998) and Cheng and Mukerjee (1998) considered the criterion of estimation capacity. According to this criterion, a regular fraction $d(A)$ of resolution three or more is evaluated on the basis of the largeness of the quantities $E_u(A)$, $u \geq 1$, where $E_u(A)$ is the number of possible models, involving all the main effect pencils and exactly u two-factor interaction pencils, which $d(A)$ can estimate when all interactions involving three or more factors are absent. The sequence $\{E_1(A), E_2(A), \ldots\}$ is called the estimation capacity sequence of $d(A)$. The computational as well as theoretical results reported by these authors indicate that the criteria of minimum aberration and estimation capacity are quite in agreement.

Observe that estimation capacity is essentially a criterion of model robustness. We refer to Paik and Federer (1973), Hedayat et al. (1974), Raktoe (1976), Monette (1983), Hedayat and Pesotan (1990), Meyer et al. (1996), and Filliben and Li (1997), among others, for results having an impact on the study of model robustness from other considerations.

The issue of blocking for regular fractions has also received attention in the literature. In the context of an m^n factorial, a blocked regular fraction $d(A, B)$ is specified by $k+s$ linearly independent pencils $\boldsymbol{a}^{(1)}, \ldots, \boldsymbol{a}^{(k)}, \boldsymbol{b}^{(1)}, \ldots, \boldsymbol{b}^{(s)}$, where $k, s \geq 1$ and $k+s < n$. It consists of m^s blocks, indexed by the $s \times 1$ vectors $\boldsymbol{\xi}$ over $GF(m)$, such that a typical block contains m^{n-k-s} distinct treatment combinations z satisfying

$$Az = \mathbf{0}, \quad Bz = \boldsymbol{\xi}, \tag{8.2.18}$$

where

$$A = (\boldsymbol{a}^{(1)}, \ldots, \boldsymbol{a}^{(k)})', \quad B = (\boldsymbol{b}^{(1)}, \ldots, \boldsymbol{b}^{(s)})'.$$

A treatment contrast is called a between block contrast in $d(A, B)$ if it involves only the treatment combinations included in $d(A, B)$, and in addition all treatment combinations belonging to the same block appear in it with the same coefficient. The defining pencils and alias sets are again dictated by (8.2.6) and (8.2.7). Then it is easy to see that given any alias set $\mathcal{B}^*$, either

$$\boldsymbol{b} \in \mathcal{M}(A', B') - \mathcal{M}(A') \qquad \text{for every } \boldsymbol{b} \in \mathcal{B}^* \tag{8.2.19}$$

or

$$\boldsymbol{b} \notin \mathcal{M}(A', B') \qquad \text{for every } \boldsymbol{b} \in \mathcal{B}^*. \tag{8.2.20}$$

Consider now any treatment contrast L belonging to a pencil $\boldsymbol{b}$ in $\mathcal{B}^*$. As before, the part of L, that involves only the treatment combinations included in $d(A, B)$, is given by $L(A)$ as specified in (8.2.10)–(8.2.12). If (8.2.19) holds, then by (8.2.12) and (8.2.18), the set $V_j(\boldsymbol{b}, A)$ is the union of m^{s-1} distinct blocks for each j. Hence then by (8.2.10)–(8.2.12), $L(A)$ is a between block contrast in $d(A, B)$. On the other hand, if (8.2.20) holds, then by (8.2.12) and (8.2.18), each set $V_j(\boldsymbol{b}, A)$ has $m^{n-k-s-1}$ treatment combinations in common with every block, so by (8.2.10)–(8.2.12), $L(A)$ is orthogonal to every between block contrast. From the points just noted, the following can be deduced:

(i) If (8.2.19) holds for any alias set $\mathcal{B}^*$, then no treatment contrast belonging to any pencil in $\mathcal{B}^*$ remains estimable in $d(A, B)$ under the presence of block effects. In this sense then the alias set $\mathcal{B}^*$ is said to be confounded with blocks.

(ii) If (8.2.20) holds for any alias set $\mathcal{B}^*$, then Lemma 8.2.5 is applicable to $\mathcal{B}^*$ even in the presence of block effects; that is, any pencil $\boldsymbol{b}$ belonging to such an alias set is estimable in $d(A, B)$ if and only if all other pencils that are aliased with $\boldsymbol{b}$ are ignorable.

By (8.2.8) and (8.2.19), the number of alias sets that are confounded with blocks equals $(m^s - 1)/(m - 1)$. Recalling that there are $(m^{n-k} - 1)/(m - 1)$ alias sets altogether, this leaves $(m^{n-k} - m^s)/(m - 1)$ alias sets that satisfy (8.2.20) and are not confounded with blocks.

Example 8.2.2. This is in continuation of Example 8.2.1. With reference to a 3^5 factorial, consider a blocked regular fraction $d(A, B)$, where A is as shown earlier in Example 8.2.1 and B consists of a single row (0, 0, 1, 0, 1). Here $m = 3$, $n = 5$, $k = 2$, $s = 1$, and $d(A, B)$ can be obtained explicitly from (8.2.18). Note that $d(A, B)$ consists of 27 distinct treatment combinations that are the same as those included in $d(A)$ of Example 8.2.1 but now arranged in three blocks of size 9 each. The defining pencils are as shown in Example 8.2.1, and none of these is a main effect pencil. Since $s = 1$, only one alias set gets confounded with blocks. By (8.2.19), this confounded alias set is given by

$$\begin{array}{lll}\{(0,0,1,0,1)', & (1,1,1,2,1)', & (1,1,2,2,2)', \\ (1,2,2,0,0)', & (1,2,0,0,1)', & (1,0,0,1,2)', \\ (1,0,1,1,0)', & (0,1,2,1,0)', & (0,1,0,1,1)'\},\end{array}$$

and it does not contain any main effect pencil. Thus the alias sets containing main effect pencils are not confounded with blocks, and the same conclusion as in Example 8.2.1 holds regarding the estimability of the main effect pencils under the absence of all interactions. ■

The reader may verify that the plan shown in Example 7.4.1 is also a blocked regular fraction with $m = 2, n = 4, k = 1, s = 2, \boldsymbol{a}^{(1)} = (1, 1, 1, 1)'$, $\boldsymbol{b}^{(1)} = (1, 1, 0, 0)'$, $\boldsymbol{b}^{(2)} = (1, 0, 1, 0)'$. We refer to Box, Hunter, and Hunter (1978) for more details on the principles underlying the issue of blocking. Very recently, several extensions of the notion of resolution and aberration to blocked regular fractions have been proposed and studied in the literature from various viewpoints. The reader is referred to Bisgaard (1994), Sun et al. (1997), Sitter et al. (1997), Mukerjee and Wu (1997b), and Chen and Cheng (1997) for these developments.

Regular fractions and extensions thereof have been studied from many other perspectives. Srivastava et al. (1984) explored parallel flats fractions that consist of treatment combinations z satisfying

$$Az \in \{\boldsymbol{c}_1, \ldots, \boldsymbol{c}_q\},$$

where A is as in (8.2.5) and $\boldsymbol{c}_1, \ldots, \boldsymbol{c}_q$ are $q(> 1)$ distinct $k \times 1$ vectors over $GF(m)$. This work was followed up by Srivastava (1987), Buhamra and Anderson (1995), Srivastava and Li (1996), and Liao et al. (1996) among others, and we refer to Ghosh (1996, sec. 7) for a review. Cheng and Li (1993) investigated the construction of regular fractions of two-level factorials when certain treatment combinations are considered infeasible and hence debarred. Several authors considered the problem of finding regular fractions when interest lies in specified factorial effects, such as the main effects and some specified two-factor interactions. In this connection we refer to Franklin (1985), Wu and Chen (1992), and Sun and Wu (1994) where more references can be found. Accounts of further group theoretic developments related to regular fractions, including those associated with a complex linear model, are available in Kobilinsky (1985), Bailey (1985), Kobilinsky and Monod (1995), and the references therein.

8.3 SEARCH DESIGNS

The theory of search linear models, due to Srivastava (1975), plays a major role in finding appropriate fractional factorial plans under model uncertainty. We begin by presenting Theorem 8.3.1 below which is central to this theory and then adopt it to our context. Consider the linear model

$$\mathbb{E}(\boldsymbol{Y}) = X_1\boldsymbol{\beta}_1 + X_2\boldsymbol{\beta}_2, \quad \mathbb{D}(\boldsymbol{Y}) = \sigma^2 I_N, \tag{8.3.1}$$

where $\boldsymbol{Y}$ is the $N \times 1$ observational vector, $X = [X_1 \vdots X_2]$ is the known design matrix, $\boldsymbol{\beta} = (\boldsymbol{\beta}_1{}', \boldsymbol{\beta}_2{}')'$ is the parametric vector, σ^2 is the common variance of the observations, and I_N is the identity matrix of order N. It is known that $\boldsymbol{\beta}_2$ can be partitioned as

$$\boldsymbol{\beta}_2 = (\boldsymbol{\beta}_{21}{}', \ldots, \boldsymbol{\beta}_{2a}{}')', \tag{8.3.2}$$

and that among $\boldsymbol{\beta}_{21}, \ldots, \boldsymbol{\beta}_{2a}$, at most k are nonnull, where k is a known positive integer that is quite small relative to a. There is, however, no knowledge about which of $\boldsymbol{\beta}_{21}, \ldots, \boldsymbol{\beta}_{2a}$ are possibly nonzero and what their values are. The objectives are to estimate $\boldsymbol{\beta}_1$ and search and estimate the possibly nonzero entities among $\boldsymbol{\beta}_{21}, \ldots, \boldsymbol{\beta}_{2a}$. If these objectives are met, then following Srivastava (1975), the design d associated with the observational vector $\boldsymbol{Y}$ in (8.3.1) is said to have resolving power $(\boldsymbol{\beta}_1, \boldsymbol{\beta}_2, k)$. Let

$$X_2 = [X_{21} \vdots \cdots \vdots X_{2a}]$$

be the partitioned form of X_2 corresponding to (8.3.2). Then the following result, due to Srivastava (1975), holds.

Theorem 8.3.1.

(a) *For the design d to have resolving power $(\boldsymbol{\beta}_1, \boldsymbol{\beta}_2, k)$ in the setup described above, it is necessary that the matrix*

$$X(i_1, \ldots, i_{2k}) = [X_1 \vdots X_{2i_1} \vdots \cdots \vdots X_{2i_{2k}}] \tag{8.3.3}$$

has full column rank for every $i_1, \ldots, i_{2k}$ $(1 \leq i_1 < \cdots < i_{2k} \leq a)$.

(b) *If in addition, $\sigma^2 = 0$, then the necessary condition stated above is also sufficient for d to have resolving power $(\boldsymbol{\beta}_1, \boldsymbol{\beta}_2, k)$.*

Proof.

(a) Let d have resolving power $(\boldsymbol{\beta}_1, \boldsymbol{\beta}_2, k)$. If possible, suppose that the stated rank condition does not hold. Then the matrix $X(i_1, \ldots, i_{2k})$ does not have full column rank for some $i_1, \ldots, i_{2k}$, say, without loss of generality, for $i_j = j (1 \leq j \leq 2k)$. Then by (8.3.3), there exist vectors $\boldsymbol{\xi}, \boldsymbol{\xi}_1, \ldots, \boldsymbol{\xi}_{2k}$, of appropriate orders and not all null, such that

$$X_1\boldsymbol{\xi} + \sum_{j=1}^{2k} X_{2j}\boldsymbol{\xi}_j = \mathbf{0}. \tag{8.3.4}$$

If $\boldsymbol{\xi}_1, \ldots, \boldsymbol{\xi}_{2k}$ are all null, then $\boldsymbol{\xi} \neq \mathbf{0}$, so X_1 is not of full column rank. But then $\boldsymbol{\beta}_1$ is not estimable, and this contradicts the fact that d has resolving power $(\boldsymbol{\beta}_1, \boldsymbol{\beta}_2, k)$. Hence $\boldsymbol{\xi}_1, \ldots, \boldsymbol{\xi}_{2k}$ are not all null. Without loss of generality, suppose that $\boldsymbol{\xi}_1 \neq \mathbf{0}$.

Consider now two models M_1 and M_2 such that according to M_1

$$\mathbb{E}(\boldsymbol{Y}) = X_1\boldsymbol{\xi}^* + \sum_{j=1}^{k} X_{2j}\boldsymbol{\xi}_j, \quad \mathbb{D}(\boldsymbol{Y}) = \sigma^2 I_N, \tag{8.3.5}$$

and under M_2,

$$\mathbb{E}(\boldsymbol{Y}) = X_1(\boldsymbol{\xi}^* - \boldsymbol{\xi}) + \sum_{j=k+1}^{2k} X_{2j}(-\boldsymbol{\xi}_j), \quad \mathbb{D}(\boldsymbol{Y}) = \sigma^2 I_N, \quad (8.3.6)$$

where $\boldsymbol{\xi}^*$ is any vector of appropriate order. By (8.3.5) and (8.3.6), the models M_1 and M_2 correspond to (8.3.1) with

$$\begin{aligned} \boldsymbol{\beta}_1 &= \boldsymbol{\xi}^*, \\ \boldsymbol{\beta}_{2j} &= \boldsymbol{\xi}_j \qquad (1 \le j \le k), \\ \boldsymbol{\beta}_{2j} &= \mathbf{0} \qquad (k+1 \le j \le a), \end{aligned}$$

and

$$\begin{aligned} &\boldsymbol{\beta}_1 = \boldsymbol{\xi}^* - \boldsymbol{\xi}, \boldsymbol{\beta}_{2j} = \mathbf{0} \qquad (1 \le j \le k \text{ and } 2k+1 \le j \le a), \\ &\boldsymbol{\beta}_{2j} = -\boldsymbol{\xi}_j \qquad (k+1 \le j \le 2k), \end{aligned}$$

respectively. Thus under both M_1 and M_2, at most k of $\boldsymbol{\beta}_{21}, \ldots, \boldsymbol{\beta}_{2a}$ are nonnull. However, the set of possibly nonnull vectors among $\boldsymbol{\beta}_{21}, \ldots, \boldsymbol{\beta}_{2a}$ under M_1 is different from that under M_2. For instance, $\boldsymbol{\beta}_{21} (= \boldsymbol{\xi}_1)$ is nonnull under M_1 but null under M_2. On the other hand, by (8.3.4)–(8.3.6), the expression for $\mathbb{E}(\boldsymbol{Y})$ under (8.3.5) is the same as that under (8.3.6). Therefore, on the basis of $\boldsymbol{Y}$, the models M_1 and M_2 are not distinguishable from each other, although the sets of possibly nonnull vectors among $\boldsymbol{\beta}_{21}, \ldots, \boldsymbol{\beta}_{2a}$ are different under these two models. This again contradicts the fact that d has resolving power $(\boldsymbol{\beta}_1, \boldsymbol{\beta}_2, k)$. Hence part (a) of the theorem follows.

(b) Let $\mathcal{C}$ be the class of subsets of $\{1, 2, \ldots, a\}$, with cardinality k or less. The empty set is also a member of $\mathcal{C}$. If $\sigma^2 = 0$, then, in view of (8.3.1) and the fact that among $\boldsymbol{\beta}_{21}, \ldots, \boldsymbol{\beta}_{2a}$ at most k are nonnull, there exists a set C in $\mathcal{C}$ and vectors $\boldsymbol{\beta}_1(C), \boldsymbol{\beta}_{2i}(C)$ $(i \in C)$, such that

$$\boldsymbol{Y} = X_1\boldsymbol{\beta}_1(C) + \sum_{i \in C} X_{2i}\boldsymbol{\beta}_{2i}(C), \quad (8.3.7)$$

with probability unity, and

$$\boldsymbol{\beta}_{2i}(C) \neq \mathbf{0} \qquad \text{for each } i \in C. \quad (8.3.8)$$

Assume that the rank condition stated in part (a) holds. In order to prove that then d has resolving power $(\boldsymbol{\beta}_1, \boldsymbol{\beta}_2, k)$, we first show that $\boldsymbol{Y}$ cannot be of the form (8.3.7), subject to (8.3.8), simultaneously for two distinct sets in $\mathcal{C}$. If possible, suppose that there are two such distinct sets, say C_1 and C_2, in $\mathcal{C}$. Then by (8.3.7) and (8.3.8),

$$X_1\boldsymbol{\beta}_1(C_1) + \sum_{i\in C_1} X_{2i}\boldsymbol{\beta}_{2i}(C_1) = X_1\boldsymbol{\beta}_1(C_2) + \sum_{i\in C_2} X_{2i}\boldsymbol{\beta}_{2i}(C_2), \tag{8.3.9}$$

where

$$\boldsymbol{\beta}_{2i}(C_1) \neq \mathbf{0} \qquad \text{for each } i \in C_1, \tag{8.3.10}$$

and

$$\boldsymbol{\beta}_{2i}(C_2) \neq \mathbf{0} \qquad \text{for each } i \in C_2. \tag{8.3.11}$$

Since the sets C_1 and C_2 are not identical, it follows from (8.3.9)–(8.3.11) that the matrix $[X_1 \,\vdots\, X_2{}^*]$, where

$$X_2{}^* = [\ldots, X_{2i}, \ldots]_{i\in C_1\cup C_2},$$

does not have full column rank. But then the assumed rank condition is violated, since $C_1 \cup C_2$ has cardinality at most $2k$. Thus we arrive at a contradiction, and it follows that $\boldsymbol{Y}$ is of the form (8.3.7), subject to (8.3.8), for a unique set C in $\mathcal{C}$. Furthermore, invoking the rank condition again, the vectors $\boldsymbol{\beta}_1(C)$ and $\boldsymbol{\beta}_{2i}(C)$ $(i \in C)$ in (8.3.7) are unique. Hence, given $\boldsymbol{Y}$, there is no ambiguity about the true model so that d has resolving power $(\boldsymbol{\beta}_1, \boldsymbol{\beta}_2, k)$. ■

Remark 8.3.1. If $\sigma^2 = 0$ and the rank condition stated in part (a) of Theorem 8.3.1 holds, then the true model can be identified with probability unity. The following considerations may then help in actually finding the unique set C and hence the true model (8.3.7). For any H in $\mathcal{C}$, let $X(H) = [X_1 \,\vdots\, X_2(H)]$ where

$$X_2(H) = [\ldots, X_{2i}, \ldots]_{i\in H}.$$

Under the rank condition, the columns of $X_2(H)$ are linearly independent. Write

$$Q(H) = X(H)\{X(H)'X(H)\}^{-1}X(H)'$$

for the orthogonal projector on the column space of $X(H)$. Also let

$$\{X(H)'X(H)\}^{-1}X(H)'\boldsymbol{Y} = \begin{bmatrix} \boldsymbol{\gamma}_1 \\ \boldsymbol{\gamma}_2(H) \end{bmatrix},$$

where

$$\boldsymbol{\gamma}_2(H) = (\ldots, \boldsymbol{\gamma}_{2i}(H)', \ldots)'_{i \in H},$$

and the orders of $\boldsymbol{\gamma}_1$ and $\boldsymbol{\gamma}_{2i}(H)$ are the same as the number of columns of X_1 and X_{2i}, respectively. Then it is not hard to see that a member H of $\mathcal{C}$ equals the unique set C, as envisaged in (8.3.7), if and only if

$$Q(H)\boldsymbol{Y} = \boldsymbol{Y} \quad \text{and} \quad \boldsymbol{\gamma}_{2i}(H) \neq \boldsymbol{0} \qquad \text{for each } i \in H,$$

in which case $\boldsymbol{\gamma}_{2i}(H)$ equals $\boldsymbol{\beta}_{2i}(C)$ for every $i \in H$. Hence one may proceed to devise an algorithm for the actual determination of the true model (8.3.7). ∎

Remark 8.3.2. For $\sigma^2 > 0$, even under the rank condition, identification of the true model is not possible with probability unity. As suggested in Srivastava (1975, 1996), the following search procedure is then helpful: Let $\mathcal{C}_0$ be a subclass of $\mathcal{C}$ consisting of those subsets of $\{1, 2, \ldots, a\}$ that have cardinality exactly k. For any $H \in \mathcal{C}_0$, let $M(H)$ be the submodel of (8.3.1) that incorporates $\boldsymbol{\beta}_1$ and $\boldsymbol{\beta}_{2i}$ $(i \in H)$ and takes $\boldsymbol{\beta}_{2i} = \boldsymbol{0}$ for $i \notin H$. For each $H \in \mathcal{C}_0$, compute

$$\text{RSS}(H) = \boldsymbol{Y}'[I_N - Q(H)]\boldsymbol{Y},$$

which is the residual sum of squares on the basis of $M(H)$. Conclude that $M(H_0)$ is the true model if $H_0(\in \mathcal{C}_0)$ is such that $\text{RSS}(H) \geq \text{RSS}(H_0)$ for every $H \in \mathcal{C}_0$. ∎

We now discuss how Theorem 8.3.1 can be applied to study fractional factorial plans under model uncertainty. We consider a general $m_1 \times \cdots \times m_n$ factorial setup, resume the use of the Kronecker notation, and return to the model (2.3.3) given by

$$\mathbb{E}(\boldsymbol{Y}) = X_d \boldsymbol{\tau}, \quad \mathbb{D}(\boldsymbol{Y}) = \sigma^2 I_N, \tag{8.3.12}$$

where, as before, $\boldsymbol{Y}$ is the $N \times 1$ observational vector under a plan d involving N runs, $\boldsymbol{\tau}$ is the vector of effects of the treatment combinations arranged in lexicographic order, and

$$X_d = ((\chi_d(u; j_1 \ldots j_n))),$$

with the indicator $\chi_d(u; j_1 \ldots j_n)$ assuming the value 1 if the uth observation according to d corresponds to the treatment combination $j_1 \ldots j_n$ and the value zero otherwise.

Consider a scenario where one wishes to use a model that incorporates the general mean and the lower-order factorial effects and possibly excludes the higher-order factorial effects but is uncertain about the status of the remaining factorial effects. To be more specific, suppose that the following information is available about the underlying model for some integers $f, t\,(0 \leq f < t \leq n)$ and $k(> 0)$:

(a) All factorial effects involving $t + 1$ or more factors are absent.

(b) Among all factorial effects involving more than f but not more than t factors, at most k are present where k is small compared to the total number of such effects, and it is unknown which of these effects are possibly present.

Because of (a) and (b), the factorial effects can be classified into three categories:

(i) Those involving more than t factors. By (a), these need not be included in the model.

(ii) Those involving more than f but not more than t factors. By (b), these have to be included in the model with the knowledge that at most k of them are actually present.

(iii) Those involving f or less factors. These, as well as the general mean, are to be included in the model.

Consideration of the intermediate category (ii), arising from (b) above, induces flexibility in the model and helps in avoiding rigid assumptions regarding the absence of specific effects in this category. The condition (b) implies that among the effects in category (ii), only a few are possibly present without spelling out what these possibly present effects are. Thus this condition signifies both effect sparsity and model uncertainty. Note that if, in particular, $t = n$, then (a) and (i) above do not arise; that is, none of the factorial effects is straightaway assumed to be absent. On the other hand, if $f = 0$, then by (b) even some main effects are potentially absent.

For ease in reference, we recall some notation from Chapter 2. Let Ω_ν be the set of binary n-tuples with at most ν components unity and the matrix P^x be as defined in (2.2.4) and (2.2.5). Also let $\boldsymbol{\beta}_x$, $\boldsymbol{\beta}^{(1)}$ and $P^{(1)}$ be as in (2.3.7)–(2.3.9). Then, in consideration of (i)–(iii) above and analogously to (2.3.12), under (a) and (b), the model (2.3.3), or equivalently (8.3.12), can be expressed as

$$\mathbb{E}(\boldsymbol{Y}) = X_d(P^{(1)})'\boldsymbol{\beta}^{(1)} + \sum\nolimits_{f,t} X_d(P^x)'\boldsymbol{\beta}_x, \qquad \mathbb{D}(\boldsymbol{Y}) = \sigma^2 I_N, \quad (8.3.13)$$

where $\sum_{f,t}$ denotes sum over $\boldsymbol{x} \in \Omega_t - \Omega_f$ and among the $\boldsymbol{\beta}_{\boldsymbol{x}}, \boldsymbol{x} \in \Omega_t - \Omega_f$, at most k are nonnull. Recall that $\boldsymbol{\beta}^{(1)}$ represents the general mean and complete sets of orthonormal contrasts belonging to factorial effects involving at most f factors. Also, for each $\boldsymbol{x} = x_1 \ldots x_n \in \Omega_t - \Omega_f$, by (2.3.7), $\boldsymbol{\beta}_{\boldsymbol{x}}$ represents a complete set of orthonormal contrasts belonging to factorial effect $F^{\boldsymbol{x}} (\equiv F_1{}^{x_1} \ldots F_n{}^{x_n})$, which obviously involves more than f but not more than t factors. The objectives are to estimate $\boldsymbol{\beta}^{(1)}$ and search and estimate the possibly nonnull $\boldsymbol{\beta}_{\boldsymbol{x}}, \boldsymbol{x} \in \Omega_t - \Omega_f$. In the spirit of Srivastava (1975), a plan d meeting these objectives will be called a search design with resolving power (f, t, k).

Comparing (8.3.1) and (8.3.13), Theorem 8.3.1 can now be paraphrased in the context of fractional factorial plans as follows:

Theorem 8.3.2.

(a) *For a plan d to be a search design with resolving power (f, t, k), it is necessary that the matrix*

$$X_d[(P^{(1)})' \vdots (P^{\boldsymbol{x}(1)})' \vdots \cdots \vdots (P^{\boldsymbol{x}(2k)})']$$

has full column rank for every choice of $2k$ distinct members $\boldsymbol{x}(1), \ldots, \boldsymbol{x}(2k)$ of $\Omega_t - \Omega_f$.

(b) *If $\sigma^2 = 0$, then the necessary condition stated above is also sufficient for d to be a search design with resolving power (f, t, k).* ■

The points noted in Remarks 8.3.1 and 8.3.2 are relevant also in connection with Theorem 8.3.2.

Example 8.3.1. Consider the setup of a 2^n factorial. For $0 \le i \le n$, let T_i be the set of treatment combinations $j_1 \ldots j_n$ such that among $j_1, \ldots, j_n$, exactly i equal 1 and the rest equal zero. Clearly, T_i has cardinality $\binom{n}{i}$. Following Srivastava and Ghosh (1976) and Srivastava and Arora (1991), consider the plans d_1 and d_2 consisting of the treatment combinations in

$$T_1 \cup T_{n-2} \cup T_{n-1} \cup T_n \quad \text{and} \quad T_0 \cup T_1 \cup T_2 \cup T_n,$$

respectively. As noted by these authors, the plans d_1 and d_2 satisfy the rank condition of Theorem 8.3.2 for (i) $f = 2, t = n, k = 1$, and (ii) $f = 1, t = n, k = 2$, respectively. Also, when $n \ge 7$, Shirakura and Tazawa (1992) observed that the plan d_3 given by the treatment combinations in $T_2 \cup T_{n-2} \cup T_n$ satisfies the same rank condition for $f = 2, t = 3$, and $k = 2$. We refer to the original papers for the proofs and other related results. ■

The issue of construction of plans satisfying the rank condition of Theorem 8.3.2 has been studied extensively in the literature. While much of this work relates to 2^n factorials, some progress has also been made with other kinds of factorials including asymmetric factorials; for example, see Anderson and Thomas (1980), Chatterjee and Mukerjee (1986a), Ghosh and Zhang (1987), and Chatterjee (1990, 1991). We refer to Gupta (1990) and Ghosh (1996) for informative reviews and further references. A comprehensive discussion, linking search linear models with other related topics, is available in Srivastava (1996).

Several authors investigated the search design problem for $\sigma^2 > 0$ and studied the probability of identifying the true model. The reader is referred to Srivastava and Mallenby (1985), Mukerjee and Chatterjee (1994), and Shirakura et al. (1996) for such results.

8.4 SUPERSATURATED DESIGNS

In an $m_1 \times \cdots \times m_n$ factorial, suppose that it is considered reasonable to assume the absence of all interactions, and further suppose that interest lies in the general mean and the main effects. Then the model corresponds to (2.3.12) with $f = t = 1$. From Theorem 2.3.2, it is known that at least

$$N_0 = 1 + \sum_{i=1}^{n} (m_i - 1) \tag{8.4.1}$$

runs are needed to keep the general mean and all the main effect contrasts estimable. A saturated plan meets this objective in exactly N_0 runs. However, particularly in exploratory studies, the number of factors, n, can be quite large and, from practical considerations, even N_0 runs may not be affordable.

One can still make some headway if there is reason to believe that among the n factors only a few are active in the sense of having nonnegligible main effects. As with search linear models, the few active factors are, however, typically not known a priori. In such situations, plans involving even less than N_0 runs may be considered with a view to estimating the general mean and identifying the nonnegligible main effects possibly via data analytic techniques like stepwise selection or ridge regression (see Lin, 1993, 1995). Plans of this kind are known as supersaturated designs.

The literature on supersaturated designs relates mostly to 2^n factorials. Hence we take $m_1 = \cdots = m_n = 2$ so that (8.4.1) reduces to $N_0 = n + 1$. Let d be a supersaturated design involving $N (< n + 1)$ runs. Then, as in (6.3.4) and (6.3.5), the information matrix under d for the parametric vector $\boldsymbol{\beta}^{(1)}$, representing the general mean and the main effects, is

$$\mathcal{I}_d = v^{-1} K_d K_d' \tag{8.4.2}$$

where $v = 2^n$ and the elements of the $(n+1) \times N$ matrix K_d are ± 1. In fact, from (6.3.1), (6.3.5), and the definition of the matrix G_d involved in (6.3.4), it is not hard to see that

$$K_d = \begin{bmatrix} 1 & 1 & \ldots & 1 \\ p_d(1,1) & p_d(1,2) & \ldots & p_d(1,N) \\ \vdots & & & \\ p_d(n,1) & p_d(n,2) & \ldots & p_d(n,N) \end{bmatrix}, \tag{8.4.3}$$

where, for $1 \le i \le n$ and $1 \le u \le N$, the quantity $p_d(i,u)$ equals $+1$ if the ith factor appears at level 1 in the uth run of d and -1 otherwise.

Suppose that the levels 0 and 1 of each factor appear equally often in d. Then d is called balanced and by (8.4.2), (8.4.3),

$$\mathcal{I}_d = v^{-1} \begin{bmatrix} N & \mathbf{0}' \\ \mathbf{0} & S_d \end{bmatrix}, \tag{8.4.4}$$

where $S_d = ((s_d(i,j))$ is an $n \times n$ matrix given by

$$s_d(i,i) = N, \qquad 1 \le i \le n,$$

$$s_d(i,j) = \sum_{u=1}^{N} p_d(i,u) p_d(j,u), \qquad 1 \le i \ne j \le n.$$

Had $s_d(i,j)$ been zero for every $i \ne j$, then by (8.4.4), $\mathcal{I}_d$ would have equaled $(N/v) I_{n+1}$ and a universal optimality result on d could have been claimed as in Lemma 2.5.2. Although the fact that $N < n+1$ rules out such a possibility (see (8.4.2)), the plan d may be anticipated to perform well in detecting the possibly present main effects provided that the quantities $s_d(i,j)$, $i \ne j$, are small in magnitude. From this consideration, Booth and Cox (1962) proposed the criterion

$$E_d(s^2) = \frac{1}{\binom{n}{2}} \sum \sum_{1 \le i < j \le n} \{s_d(i,j)\}^2$$

as a performance characteristic of a balanced supersaturated design. Several other criteria for choosing supersaturated designs are discussed in Li and Wu (1997) where a brief review of the literature is also available.

Algebraic as well as algorithmic methods for constructing supersaturated designs, which minimize $E_d(s^2)$ or behave well with respect to other criteria, have been investigated by several authors. The reader is referred to Lin (1993, 1995), Wu (1993), Nguyen (1996a), Li and Wu (1997), Tang and Wu (1997), and Cheng (1997) for the details of such constructions. More references are available in Cheng (1998). We also refer to Gupta and Chatterjee (1998) for a review of supersaturated designs highlighting their relationship with search designs, and to Chatterjee and Gupta (1997) for some results on supersaturated designs for general m^n symmetric factorials.

The idea of supersaturated designs depends critically on effect sparsity in the form of the supposition that only a few of the n factors are actually active. The same consideration has led to the study of another phenomenon, namely projectivity. Following Box and Tyssedal (1996), a fractional factorial plan involving n factors has projectivity $n^*(< n)$ if for every choice of n^* factors, it produces a complete factorial. As before, suppose that only a few factors are active, though it is not known a priori what these active factors are. One may then wish to identify the active factors via an initial analysis, based on a model incorporating only the general mean and the main effects, and then proceed to explore a model that involves only the active factors so identified and interactions among them. Since the active factors are not known a priori, the feasibility of such an analysis, which possibly exploits data analytic techniques, depends on the projectivity of the plan used. Several authors, like Lin and Draper (1992), Cheng (1995), Wang and Wu (1995), and Box and Tyssedal (1996), examined fractional factorial plans based on orthogonal arrays with regard to projectivity and related issues. We refer to Cheng (1998) for a review of the work in this area together with a discussion on the connection with search designs.

In concluding this chapter, we emphasize again that our aim here was to familiarize the reader with only the fundamentals of minimum aberration, search, and supersaturated designs. It is hoped that this introduction, together with the references cited, will help in pursuing further studies on these emerging areas.

EXERCISES

8.1. Let

$$d = \{00000, 01100, 00011, 01111, 11010, 10110, 11001, 10101\}$$

be a regular fraction of a 2^5 factorial.

(a) Show that the defining pencils for d are $(1, 1, 1, 0, 0)'$, $(1, 0, 0, 1, 1)'$, and $(0, 1, 1, 1, 1)'$.

(b) Find the alias sets with reference to d.

(c) Is it possible to partition d into two blocks, each consisting of four treatment combinations, retaining estimability of all main effect pencils under the absence of all interactions?

8.2. **(a)** Let $m(\geq 2)$ be a prime or prime power and $r(\geq 3)$ be an integer. Consider a regular fraction of an m^n factorial in m^r runs. If such a fraction has resolution four, then use Theorem 8.2.2 and Remark 2.6.3 to show that

$$n \leq \frac{m^{r-1} - 1}{m - 1} + 1.$$

(b) Show that the above upper bound is attainable if $m = 2$. [*Hint*: Consider $r \times 1$ vectors over $GF(2)$ having an odd number of 1's, as columns of the matrix C in Theorem 8.2.2.]

(c) Also show that the upper bound in part (a) is not necessarily attainable if $m \geq 3$. [*Hint*: See Remark 3.4.1.]

8.3. For a 2^7 factorial, give examples of two regular fractions, each involving 32 runs, such that both the fractions have resolution four but one of them has less aberration than the other. In this setup, does there exist any regular fraction of resolution greater than four?

8.4. Verify the contents of Remark 8.3.1.

Appendix

Several construction procedures described in this book are based on Hadamard matrices and difference matrices. In Section A.1 we indicate the construction of Hadamard matrices of orders less than 50. Some small difference matrices are presented in Section A.2. Finally a guide to orthogonal arrays of strength two with at most 50 rows is given in Section A.3.

A.1 HADAMARD MATRICES

We indicate the construction of Hadamard matrices of orders less than 50. As before, a Hadamard matrix of order N will be denoted by H_N. In addition to Lemma 3.2.1, according to which the Kronecker product of two Hadamard matrices is also a Hadamard matrix, the following construction procedure will be helpful:

Let N be an integral multiple of 4 such that $N - 1 (= w)$ is a prime or a prime power. Denote the elements of $GF(w)$ by $\rho_0 = 0$, $\rho_1 = 1$, $\rho_2 = y, \ldots, \rho_{w-1} = y^{w-2}$, where y is a primitive element. For $0 \le j \le w - 1$, define the sets

$$S_j = \{\rho_j + \rho_1, \rho_j + \rho_3, \ldots, \rho_j + \rho_{w-2}\}.$$

Let $\mathbf{1}$ denote the $w \times 1$ vector with all elements unity, and

$$A = ((a_{ij}))_{i,j=0,1,\ldots,w-1},$$

where a_{ij} equals $+1$ if $\rho_i \in S_j$ and -1 otherwise. Then H_N can be constructed as

$$H_N = \begin{bmatrix} 1 & \mathbf{1}' \\ \mathbf{1} & A \end{bmatrix}. \tag{A.1.1}$$

Observe that once S_0 is obtained, the subsequent steps in the above construction follow quite routinely. In particular, for any $j (1 \le j \le w-1)$, S_j can be found simply by adding ρ_j to each element of S_0.

Hadamard Matrices ($2 \le N < 50$)

1. $H_2 = \begin{bmatrix} 1 & 1 \\ 1 & -1 \end{bmatrix}$.
2. $H_4 = H_2 \otimes H_2$.
3. $H_8 = H_4 \otimes H_2$.
4. H_{12}: Use (A.1.1). Start with $S_0 = \{1, 2^2, 2^4, 2^6, 2^8\}$, where each element is reduced modulo 11. Note that 2 is a primitive element of $GF(11)$.
5. $H_{16} = H_8 \otimes H_2$.
6. H_{20}: Use (A.1.1). Start with $S_0 = \{1, 2^2, 2^4, \ldots, 2^{16}\}$, where each element is reduced modulo 19. Note that 2 is a primitive element of $GF(19)$.
7. $H_{24} = H_{12} \otimes H_2$.
8. H_{28}: Use (A.1.1). Start with $S_0 = \{1, y^2, y^4, \ldots, y^{24}\}$, where y is a primitive element of $GF(27)$ and $y^3 = y + 2$. Thus

$$S_0 = \{1, y^2, y^2+2y, y^2+y+1, 2y^2+2, y^2+y, y^2+2, 2y, \\ 2y+1, y^2+2y+1, 2y^2+y+1, 2y+2, 2y^2+2y+1\}.$$

9. $H_{32} = H_{16} \otimes H_2$.
10. $H_{36} = \begin{bmatrix} 1 & -\mathbf{1}' \\ \mathbf{1} & A \end{bmatrix}$, where $\mathbf{1}$ is a 35×1 vector with each element unity and

$$A = \begin{bmatrix} a_1 & a_2 & \cdots & a_{34} & a_{35} \\ a_{35} & a_1 & \cdots & a_{33} & a_{34} \\ \vdots & & & & \\ a_2 & a_3 & \cdots & a_{35} & a_1 \end{bmatrix},$$

the first row of A being given by

$$-, +, -, +, +, +, -, -, -, +, +, +, +, +, -, +, +,$$
$$+, -, -, +, -, -, -, -, +, -, +, -, +, +, -, -, +, -,$$

where $+$ and $-$ respectively stand for $+1$ and -1.

11. $H_{40} = H_{20} \otimes H_2$.

12. H_{44}: Use (A.1.1). Start with $S_0 = \{1, 3^2, 3^4, \ldots, 3^{40}\}$, where each element is reduced modulo 43. Note that 3 is a primitive element of $GF(43)$.

13. $H_{48} = H_{24} \otimes H_2$.

For some of the above cases, other nonisomorphic solutions for H_N may exist.

A.2 DIFFERENCE MATRICES

We provide a guide to some small difference matrices. As noted in Section 3.5, a difference matrix $D(N, N; 2)$ is equivalent to a Hadamard matrix of order N. Hence, in view of the contents of the previous section, such difference matrices are left out of consideration here. Similarly, by virtue of Remark 3.5.1, we find it redundant to consider here difference matrices $D(m, m; m)$ where m is a prime or a prime power.

1. $D(6, 6; 3)$: See Example 3.5.3.
2. $D(8, 8; 4)$: See Example 3.5.2.
3. $D(9, 9; 3)$: Use Lemma 3.5.1 with $\lambda = m = 3$.
4. $D(10, 10; 5)$: Use Lemma 3.5.2 with $m = 5$.
5. From Seiden (1954), we have $D(12, 12; 3)$ as displayed below. See also Remark 3.5.3.

$$D(12, 12; 3) = \begin{bmatrix} 0 & 0 & 0 & 1 & 1 & 0 & 0 & 1 & 0 & 2 & 2 & 0 \\ 0 & 0 & 0 & 0 & 2 & 0 & 2 & 0 & 2 & 0 & 0 & 1 \\ 0 & 0 & 1 & 0 & 0 & 2 & 1 & 2 & 0 & 0 & 1 & 0 \\ 0 & 0 & 2 & 2 & 0 & 1 & 0 & 0 & 1 & 1 & 0 & 0 \\ 0 & 1 & 2 & 2 & 0 & 0 & 1 & 1 & 2 & 0 & 2 & 2 \\ 0 & 1 & 2 & 1 & 2 & 1 & 2 & 2 & 2 & 2 & 1 & 0 \\ 0 & 1 & 0 & 0 & 2 & 2 & 0 & 2 & 1 & 1 & 2 & 2 \\ 0 & 1 & 1 & 2 & 1 & 2 & 2 & 0 & 0 & 2 & 0 & 2 \\ 0 & 2 & 1 & 2 & 1 & 0 & 0 & 2 & 2 & 1 & 1 & 1 \\ 0 & 2 & 1 & 0 & 0 & 1 & 2 & 1 & 1 & 2 & 2 & 1 \\ 0 & 2 & 2 & 1 & 2 & 2 & 1 & 1 & 0 & 1 & 0 & 1 \\ 0 & 2 & 0 & 1 & 1 & 1 & 1 & 0 & 1 & 0 & 1 & 2 \end{bmatrix}.$$

6. From Seberry (1979), we have $D(12, 12; 4)$ as displayed below. Here, 00, 01, 10, 11 stand, respectively, for the elements 0, 1, y, $y + 1$ of

$GF(4)$ with y denoting a primitive element:

$$D(12, 12; 4) = \begin{bmatrix} 00 & 00 & 00 & 00 & 00 & 00 & 00 & 00 & 00 & 00 & 00 & 00 \\ 00 & 00 & 00 & 01 & 01 & 01 & 11 & 11 & 11 & 10 & 10 & 10 \\ 00 & 00 & 00 & 11 & 11 & 11 & 10 & 10 & 10 & 01 & 01 & 01 \\ 00 & 11 & 01 & 10 & 01 & 11 & 01 & 10 & 00 & 11 & 00 & 10 \\ 00 & 11 & 01 & 11 & 10 & 01 & 00 & 01 & 10 & 10 & 11 & 00 \\ 00 & 11 & 01 & 01 & 11 & 10 & 10 & 00 & 01 & 00 & 10 & 11 \\ 00 & 01 & 10 & 11 & 00 & 10 & 01 & 00 & 11 & 01 & 11 & 10 \\ 00 & 01 & 10 & 10 & 11 & 00 & 11 & 01 & 00 & 10 & 01 & 11 \\ 00 & 01 & 10 & 00 & 10 & 11 & 00 & 11 & 01 & 11 & 10 & 01 \\ 00 & 10 & 11 & 01 & 10 & 00 & 01 & 11 & 10 & 01 & 00 & 11 \\ 00 & 10 & 11 & 00 & 01 & 10 & 10 & 01 & 11 & 11 & 01 & 00 \\ 00 & 10 & 11 & 10 & 00 & 01 & 11 & 10 & 01 & 00 & 11 & 01 \end{bmatrix}.$$

7. From Wang and Wu (1991), we have $D(12, 6; 6)$ as displayed below:

$$D(12, 6; 6) = \begin{bmatrix} 0 & 0 & 0 & 0 & 0 & 0 \\ 0 & 1 & 3 & 2 & 4 & 0 \\ 0 & 2 & 0 & 1 & 5 & 2 \\ 0 & 3 & 1 & 5 & 4 & 2 \\ 0 & 4 & 3 & 5 & 2 & 1 \\ 0 & 5 & 5 & 3 & 1 & 1 \\ 0 & 0 & 2 & 3 & 2 & 3 \\ 0 & 1 & 2 & 4 & 0 & 5 \\ 0 & 2 & 5 & 2 & 3 & 4 \\ 0 & 3 & 4 & 1 & 1 & 4 \\ 0 & 4 & 1 & 0 & 3 & 5 \\ 0 & 5 & 4 & 4 & 5 & 3 \end{bmatrix}.$$

8. $D(14, 14; 7)$: Use Lemma 3.5.2 with $m = 7$.

A.3 SELECTED ORTHOGONAL ARRAYS

We provide a guide to the construction of selected orthogonal arrays having strength two and up to 50 rows. For notational simplicity, an $OA(N, n, m_1 \times \cdots \times m_n, 2)$ will be denoted by $L_N(m_1 \times \cdots \times m_n)$. Trivially, deletion of one or more column(s) of any array indicated below will produce more orthogonal arrays. In the sequel, whenever needed, we will use the arrays of the form $L_{m_1 m_2}(m_1 \times m_2)$, given simply by the rows $j_1 j_2 (0 \le j_1 \le m_1 - 1; 0 \le j_2 \le m_2 - 1)$.

1. $L_4(2^3)$: Use Theorem 3.2.2.
2. $L_8(2^7)$: Use Theorem 3.2.2.
3. $L_8(4 \times 2^4)$: See Example 4.3.1.
4. $L_9(3^4)$: See Example 3.4.2.
5. $L_{12}(2^{11})$: Use Theorem 3.2.2.
6. $L_{12}(2^4 \times 3)$: See Wang and Wu (1991).
7. $L_{12}(2^2 \times 6)$: Following Section 4.4, construct as

$$[(0, 1)' \odot D \vdots (0, 0)' \odot (0, 1, 2, 3, 4, 5)'],$$

   where $\odot$ denotes Kronecker sum and

$$D = \begin{bmatrix} 0 & 0 & 0 & 0 & 0 & 0 \\ 0 & 0 & 0 & 1 & 1 & 1 \end{bmatrix}'.$$

8. $L_{16}(4^5)$: Use Theorem 3.4.1 with $m = 4$, $g = 2$.
9. $L_{16}(4^{5-u} \times 2^{3u})$, $1 \le u \le 5$: Choose any u columns of $L_{16}(4^5)$ and replace each of these by three 2-symbol columns as in Example 4.2.2.
10. $L_{16}(2^8 \times 8)$: Following Section 4.4, construct as

$$[(0, 1)' \odot D(8, 8; 2) \vdots (0, 0)' \odot (0, 1, \ldots, 7)'].$$

11. $L_{18}(2 \times 3^7)$: See Remark 5.4.3.
12. $L_{18}(3^6 \times 6)$: See Example 4.5.2.
13. $L_{20}(2^{19})$: Use Theorem 3.2.2.
14. $L_{20}(2^8 \times 5)$: See Wang and Wu (1991).
15. $L_{24}(2^{23})$: Use Theorem 3.2.2.
16. $L_{24}(2^{16} \times 3)$: Following Section 4.4, construct as

$$[(0, 1)' \odot D(12, 12; 2) \vdots (0, 0)' \odot L_{12}(2^4 \times 3)].$$

17. $L_{24}(4 \times 2^{20})$: Use Theorem 4.3.1 with $N = 12$, $T = 2$.
18. $L_{24}(2^{14} \times 6)$: Construct as in 16 above with $L_{12}(2^4 \times 3)$ there replaced by $L_{12}(2^2 \times 6)$.
19. $L_{24}(4 \times 2^{13} \times 3)$: Use Theorem 4.4.2 with $N = 12$, $T = 2$, $s = 1$, and G given by $L_{12}(2^4 \times 3)$; see also Remark 4.4.1.
20. $L_{24}(6 \times 4 \times 2^{11})$: See Example 4.3.2.
21. $L_{25}(5^6)$: Use Theorem 3.4.1 with $m = 5$, $g = 2$.

22. $L_{27}(3^{13})$: Use Theorem 3.4.3 with $m = r = 3$.

23. $L_{27}(9 \times 3^9)$: See Example 4.6.1.

24. $L_{28}(2^{27})$: Use Theorem 3.2.2.

25. $L_{28}(2^{12} \times 7)$: See Suen (1989b) and Dey and Midha (1996).

26. $L_{32}(4^9 \times 2^4)$: Following Section 4.4, construct as

$$[(0, 1, y, y + 1)' \odot D(8, 8; 4) \vdots (0, 0, 0, 0)' \odot L_8(4 \times 2^4)],$$

where y is a primitive element of $GF(4)$.

27. $L_{32}(4^8 \times 8)$: Construct as in 26 above with $L_8(4 \times 2^4)$ there replaced by $(0, 1, 2, \ldots, 7)'$.

28. $L_{32}(4^{9-u} \times 2^{4+3u})$, $1 \leq u \leq 9$: Choose any u of the 4-symbol columns of $L_{32}(4^9 \times 2^4)$ and replace each of these by three 2-symbol columns as in Example 4.2.2.

29. $L_{32}(4^{8-u} \times 2^{3u} \times 8)$, $1 \leq u \leq 8$: Start from 27 above and use the technique of replacement as in 28 above.

30. $L_{36}(2^{35})$: Use Theorem 3.2.2.

31. $L_{36}(3^{13} \times 2^4)$: Following Section 4.4, construct as

$$[(0, 1, 2)' \odot D(12, 12; 3) \vdots (0, 0, 0)' \odot L_{12}(3 \times 2^4)].$$

32. $L_{36}(3^{12} \times 2^{11})$: Construct as in 31 above with $L_{12}(3 \times 2^4)$ there replaced by $L_{12}(2^{11})$.

33. $L_{36}(3^{12} \times 2^2 \times 6)$: Construct as in 31 above with $L_{12}(3 \times 2^4)$ there replaced by $L_{12}(2^2 \times 6)$.

34. $L_{36}(3^{13} \times 4)$: Construct as in 31 above with $L_{12}(3 \times 2^4)$ there replaced by $L_{12}(3 \times 4)$.

35. $L_{36}(6^3 \times 3^7)$: See Finney (1982).

36. $L_{36}(2^{13} \times 9)$: See Suen (1989b) and Dey and Midha (1996).

37. $L_{36}(2^{13} \times 3^4)$: Replace the symbols in the 9-symbol column of $L_{36}(2^{13} \times 9)$ by the rows of $L_9(3^4)$ as discussed in Section 4.2.

38. $L_{40}(2^{39})$: Use Theorem 3.2.2.

39. $L_{40}(4 \times 2^{36})$: Use Theorem 4.3.1 with $N = 20$, $T = 2$.

40. $L_{40}(2^{28} \times 5)$: Following Section 4.4, construct as

$$[(0, 1)' \odot D(20, 20; 2) \vdots (0, 0)' \odot L_{20}(2^8 \times 5)].$$

41. $L_{40}(4 \times 2^{25} \times 5)$: Use Theorem 4.4.2 with $N = 20$, $T = 2$, $s = 1$ and G given by $L_{20}(2^8 \times 5)$; see also Remark 4.4.1.

42. $L_{44}(2^{43})$: Use Theorem 3.2.2.

43. $L_{45}(3^8 \times 5)$: Following Section 4.4, construct as

$$[(0, 1, 2)' \odot D(15, 7; 3) \vdots (0, 0, 0)' \odot L_{15}(3 \times 5)].$$

See Hedayat et al. (1996) for more information on $D(15, 7; 3)$.

44. $L_{48}(4^{12} \times 2^{11})$: Following Section 4.4, construct as

$$[(0, 1, y, y + 1)' \odot D(12, 12; 4) \vdots (0, 0, 0, 0)' \odot L_{12}(2^{11})],$$

where y is a primitive element of $GF(4)$.

45. $L_{48}(4^{12} \times 2^4 \times 3)$: Construct as in 44 above with $L_{12}(2^{11})$ there replaced by $L_{12}(2^4 \times 3)$.

46. $L_{48}(4^{12} \times 2^2 \times 6)$: Construct as in 44 above with $L_{12}(2^{11})$ there replaced by $L_{12}(2^2 \times 6)$.

47. $L_{48}(4^{12-u} \times 2^{11+3u})$, $1 \leq u \leq 12$: Choose any u of the 4-symbol columns of $L_{48}(4^{12} \times 2^{11})$ and replace each of these by three 2-symbol columns as in Example 4.2.2.

48. $L_{48}(4^{12-u} \times 2^{4+3u} \times 3)$, $1 \leq u \leq 12$: Start from 45 above and use the technique of replacement as employed in 47 above.

49. $L_{48}(4^{12-u} \times 2^{2+3u} \times 6)$, $1 \leq u \leq 12$: Start from 46 above and use the technique of replacement as employed in 47 above.

50. $L_{48}(4^{13} \times 3)$: Construct as in 44 above with $L_{12}(2^{11})$ there replaced by $L_{12}(4 \times 3)$.

51. $L_{48}(8 \times 2^{40})$: Use Theorem 4.4.4 with $N = 12$, $T = 4$ and start with $L_{12}(2^{11})$.

52. $L_{48}(8 \times 2^{33} \times 3)$: Use Theorem 4.4.4 with $N = 12$, $T = 4$ and start with $L_{12}(2^4 \times 3)$.

53. $L_{48}(8 \times 2^{31} \times 6)$: Use Theorem 4.4.4 with $N = 12$, $T = 4$ and start with $L_{12}(2^2 \times 6)$.

54. $L_{49}(7^8)$: Use Theorem 3.4.1 with $m = 7$, $g = 2$.

55. $L_{50}(5^{11} \times 2)$: Following Section 4.4, construct as

$$[(0, 1, 2, 3, 4)' \odot D(10, 10; 5) \vdots (0, 0, 0, 0, 0)' \odot L_{10}(5 \times 2)].$$

References

Addelman, S. (1962a). Orthogonal main effect plans for asymmetrical factorial experiments. *Technometrics* **4**, 21–46.

Addelman, S. (1962b). Symmetrical and asymmetrical fractional factorial plans. *Technometrics* **4**, 47–58.

Addelman, S. (1972). Recent developments in the design of factorial experiments. *J. Amer. Statist. Assoc.* **67**, 103–111.

Addelman, S. and O. Kempthorne (1961a). *Orthogonal Main Effect Plans*. Aerospace Res. Lab. Tech. Rep. 79.

Addelman, S. and O. Kempthorne (1961b). Some main effect plans and orthogonal arrays of strength two. *Ann. Math. Statist.* **32**, 1167–1176.

Adhikary, B. and P. Das (1990). A study of efficiency of proportional frequency plans. *Sankhyā* **B52**, 338–342.

Agrawal, V. and A. Dey (1982). A note on orthogonal main effect plans for asymmetrical factorials. *Sankhyā* **B44**, 278–282.

Agrawal, V. and A. Dey (1983). Orthogonal Resolution IV designs for some asymmetrical factorials. *Technometrics* **25**, 197–199.

Anderson, D. A. and W. T. Federer (1995). Representation and bounds for the general fractional factorial. *Commun. Statist. Theory Meth.* **24**, 363–373.

Anderson, D. A. and A. M. Thomas (1980). Weakly resolvable IV.3 search designs for the p^n factorial experiment. *J. Statist. Plann. Inference* **4**, 299–312.

Atkinson, A. C. and A. N. Donev (1996). Experimental designs optimally balanced for trend. *Technometrics* **38**, 333–341.

Bagiatis, C. (1990). E-optimal 2^k saturated designs. *Statistics* **21**, 35–44.

Bailey, R. A. (1977). Patterns of confounding in factorial designs. *Biometrika* **64**, 597–603.

Bailey, R. A. (1985). Factorial designs and Abelian groups. *Linear Algebra Appl.* **70**, 349–368.

Bailey, R. A., C.-S. Cheng, and P. Kipnis (1992). Construction of trend-resistant factorial designs. *Statist. Sinica* **2**, 393–411.

Balasubramanian, K. and A. Dey (1996). D-optimal designs with minimal and nearly minimal number of units. *J. Statist. Plann. Inference* **52**, 255–262.

Banerjee, K. S. (1975). *Weighing Designs: For Chemistry, Medicine, Economics, Operations Research, Statistics*. New York: Marcel Dekker.

Bapat, R. B. and A. Dey (1991). Optimal block designs with minimal number of observations. *Statist. Probab. Lett.* **11**, 399–402.

Baumert, L. D., S. W. Golomb, and M. Hall Jr. (1962). Discovery of an Hadamard matrix of order 92. *Bull. Amer. Math. Soc.* **68**, 237–238.

Birkes, D. and Y. Dodge (1991). Optimal $a \times b$ connected designs with $a + b$ observations. *J. Statist. Plann. Inference* **28**, 49–59.

Bisgaard, S. (1994). A note on the definition of resolution for blocked 2^{k-p} designs. *Technometrics* **36**, 308–311.

Booth, K. H. V. and D. R. Cox (1962). Some systematic supersaturated designs. *Technometrics* **4**, 489–495.

Bose, R. C. (1947). Mathematical theory of the symmetrical factorial design. *Sankhyā* **8**, 107–166.

Bose, R. C. and K. A. Bush (1952). Orthogonal arrays of strength two and three. *Ann. Math. Statist.* **23**, 508–524.

Box, G. E. P. and J. S. Hunter (1961a). The 2^{k-p} fractional factorial designs. *Technometrics* **3**, 311–352.

Box, G. E. P. and J. S. Hunter (1961b). The 2^{k-p} fractional factorial designs, II. *Technometrics* **3**, 449–458.

Box, G. E. P., W. G. Hunter, and J. S. Hunter (1978). *Statistics for Experimenters*. New York: Wiley.

Box, G. E. P. and J. Tyssedal (1996). Projective properties of certain orthogonal arrays. *Biometrika* **83**, 950–955.

Box, G. E. P. and K. B. Wilson (1951). On the experimental attainment of optimum conditions. *J. Roy. Statist. Soc.* **B13**, 1–45.

Buhamra, S. and D. A. Anderson (1995). The spectrum of the information matrix for s^n parallel flats fractions when s is a prime. *J. Statist. Plann. Inference* **47**, 333–345.

Burgess, L. and D. J. Street (1994). Algorithms for constructing orthogonal main effect plans. *Utilitas Math.* **46**, 33–48.

Bush, K. A. (1952). Orthogonal arrays of index unity. *Ann. Math. Statist.* **23**, 426–434.

Chacko, A. and A. Dey (1981). Some orthogonal main effect plans for asymmetrical factorials. *Sankhyā* **B43**, 384–391.

Chacko, A., A. Dey, and G. V. S. Ramakrishna (1979). Orthogonal main effect plans for asymmetrical factorials. *Technometrics* **21**, 269–270.

Chadjiconstantinidis, S., C.-S. Cheng, and C. Moyssiadis (1989). Construction of optimal fractional factorial resolution V designs with $N \equiv 2 \bmod 16$ observations. *J. Statist. Plann. Inference* **23**, 153–161.

Chadjipantelis, T. and S. Kounias (1985). Supplementary difference sets and D-optimal designs for $n \equiv 2 (\mathrm{mod}\ 4)$. *Discrete Math.* **57**, 211–216.

Chakrabarti, M. C. (1962). *Mathematics of Design and Analysis of Experiments*. Bombay: Asia Publishing.

Chakravarti, I. M. (1956). Fractional replication in asymmetrical factorial designs and partially balanced arrays. *Sankhyā* **17**, 143–164.

Chakravarty, R. and A. Dey (1976). On the construction of balanced and orthogonal arrays. *Canadian J. Statist.* **4**, 109–117.

Chatterjee, K. (1990). Search designs for searching for one among the two- and three-factor interaction effects in the general symmetric and asymmetric factorials. *Ann. Inst. Statist. Math.* **42**, 783–803.

Chatterjee, K. (1991). Search designs for searching three-factor interaction effects in the general symmetric and asymmetric factorials. *Sankhyā* **B53**, 304–326.

Chatterjee, K. and S. Gupta (1997). Construction of supersaturated designs involving s-level factors. Preprint.

Chatterjee, K. and R. Mukerjee (1986a). Some search designs for symmetric and asymmetric factorials. *J. Statist. Plann. Inference* **13**, 357–363.

Chatterjee, K. and R. Mukerjee (1986b). Linear trend-free orthogonal main effect plans. *Rep. Statist. Appl. Res., JUSE* **33**, 1–8.

Chatterjee, K. and R. Mukerjee (1993). D-optimal saturated main effect plans for $2 \times s_2 \times s_3$ factorials. *J. Combin. Inf. System Sc.* **18**, 116–122.

Chatterjee, S. K. (1982). Some recent developments in the theory of asymmetric factorial experiments—A review. *Sankhyā* **A44**, 103–113.

Chen, H. and C.-S. Cheng (1997). Theory of optimal blocking of 2^{n-m} designs. Preprint.

Chen, H. and A. S. Hedayat (1996). 2^{n-l} designs with weak minimum aberration. *Ann. Statist.* **24**, 2536–2548.

Chen, J. (1992). Some results on 2^{n-k} fractional factorial designs and search for minimum aberration designs. *Ann. Statist.* **20**, 2124–2141.

Chen, J., D. X. Sun, and C. F. J. Wu (1993). A catalogue of two-level and three-level fractional factorial designs with small runs. *Internat. Statist. Rev.* **61**, 131–145.

Chen, J. and C. F. J. Wu (1991). Some results on s^{n-k} fractional factorial designs with minimum aberration or optimal moments. *Ann. Statist.* **19**, 1028–1041.

Cheng, C.-S. (1978). Optimality of certain asymmetrical experimental designs. *Ann. Statist.* **6**, 1239–1261.

Cheng, C.-S. (1980a). Optimality of some weighing and 2^n fractional factorial designs. *Ann. Statist.* **8**, 436–446.

Cheng, C.-S. (1980b). Orthogonal arrays with variable numbers of symbols. *Ann. Statist.* **8**, 447–453.

Cheng, C.-S. (1985). Run orders of factorial designs. In *Proc. Berkeley Conf. in honor of J. Neyman and J. Kiefer*, vol. 2 (L. M. Le Cam and R. A. Olshen, eds.), pp. 619–633, Belmont, CA: Wadsworth.

Cheng, C.-S. (1989). Some orthogonal main effect plans for asymmetrical factorials. *Technometrics* **31**, 475–477.

Cheng, C.-S. (1990). Construction of run orders of factorial designs. In *Statistical Design and Analysis of Industrial Experiments* (S. Ghosh, ed.), pp. 423–439, New York: Marcel Dekker.

Cheng, C.-S. (1995). Some projection properties of orthogonal arrays. *Ann. Statist.* **23**, 1223–1233.

Cheng, C.-S. (1997). $E(s^2)$ optimal supersaturated designs. *Statist. Sinica* **7**, 929–939.

Cheng, C.-S. (1998). Projectivity and resolving power. *J. Combin. Inf. System Sc.*, to appear.

Cheng, C.-S. and M. Jacroux (1988). The construction of trend-free run orders of two-level factorial designs. *J. Amer. Statist. Assoc.* **83**, 1152–1158.

Cheng, C.-S. and C. C. Li (1993). Constructing orthogonal fractional factorial designs when some factor level combinations are debarred. *Technometrics* **35**, 277–283.

Cheng, C.-S., J. C. Masaro, and C. S. Wong (1985). Optimal weighing designs. *SIAM J. Alg. Disc. Meth.* **6**, 259–267.

Cheng, C.-S. and R. Mukerjee (1998). Regular fractional factorial designs with minimum aberration and maximum estimation capacity. *Ann. Statist.*, to appear.

Cheng, C.-S. and D. M. Steinberg (1991). Trend robust two-level factorial designs. *Biometrika* **78**, 325–336.

Cheng, C.-S., D. M. Steinberg, and D. X. Sun (1998). Minimum aberration and model robustness. *J. Roy. Statist. Soc.* **B**, to appear.

Collombier, D. (1988). Optimality of some fractional factorial designs. In *Optimal Design and Analysis of Experiments* (Y. Dodge et al., eds.), pp. 39–45, Amsterdam: North-Holland.

Collombier, D. (1992). Generally optimal main-effect fractions of $u \times v$ designs with $u + v$ units. *Comput. Statist. Data Anal.* **14**, 333–342.

Coster, D. C. (1993a). Trend-free run orders of mixed-level fractional factorial designs. *Ann. Statist.* **21**, 2072–2086.

Coster, D. C. (1993b). Tables of minimum cost, linear trend-free run sequences for two- and three-level fractional factorial design. *Comput. Statist. Data Anal.* **16**, 325–336.

Coster, D. C. and C.-S. Cheng (1988). Minimum cost trend-free run orders of fractional factorial designs. *Ann. Statist.* **16**, 1188–1205.

Cox, D. R. (1958). *Planning of Experiments*. New York: Wiley.

Daniel, C. and F. Wilcoxon (1966). Fractional 2^{p-q} plans robust against linear and quadratic trends. *Technometrics* **8**, 259–278.

Dawson, J. E. (1985). A construction for generalized Hadamard matrices $GH(4q, EA(q))$. *J. Statist. Plann. Inference* **11**, 103–110.

de Launey, W. (1986). A survey of generalized Hadamard matrices and difference matrices $D(k, \lambda; G)$ with large k. *Utilitas Math.* **30**, 5–29.

Delsarte, P. (1973). An algebraic approach to the association schemes of coding theory. *Phillips Res. Rep. Suppl.* **10**.

Dey, A. (1985). *Orthogonal Fractional Factorial Designs*. New York: Halsted Press.

Dey, A. (1986). *Theory of Block Designs*. New York: Halsted Press.

Dey, A. (1993). Some orthogonal arrays with variable symbols. *J. Combin. Inf. System Sc.* **18**, 209–215.

Dey, A. and V. Agrawal (1985). Orthogonal fractional plans for asymmetrical factorials derivable from orthogonal arrays. *Sankhyā* **B47**, 56–66.

Dey, A. and C. K. Midha (1996). Construction of some asymmetrical orthogonal arrays. *Statist. Probab. Lett.* **28**, 211–217.

Dey, A. and C. K. Midha (1998). Addition or deletion? *Statist. Probab. Lett.* **37**, 409–414.

Dey, A. and R. Mukerjee (1998). Techniques for constructing asymmetric orthogonal arrays. *J. Combin. Inf. System Sc.*, to appear.

Dey, A. and G. V. S. Ramakrishna (1977). A note on orthogonal main effect plans. *Technometrics* **19**, 511–512.

Dey, A., K. R. Shah, and A. Das (1995). Optimal block designs with minimal and nearly minimal number of units. *Statist. Sinica* **5**, 547–558.

Dickinson, A. W. (1974). Some run orders requiring a minimum number of factor level changes for 2^4 and 2^5 main effect plans. *Technometrics* **16**, 31–37.

Dong, F. (1993). On the identification of active contrasts in unreplicated fractional factorials. *Statist. Sinica* **3**, 209–217.

Draper, N. R. and D. K. J. Lin (1990). Capacity considerations for two-level fractional factorial designs. *J. Statist. Plann. Inference* **24**, 25–35 (Correction: *ibid.* **25**, 205).

Draper, N. R. and T. J. Mitchell (1968). Construction of the set of 256-run designs of resolution ≥ 5 and the set of even 512-run designs of resolution ≥ 6 with special reference to the unique saturated designs. *Ann. Math. Statist.* **39**, 246–255.

Draper, N. R. and D. M. Stoneman (1968). Factor changes and linear trends in eight-run two-level factorial designs. *Technometrics* **10**, 301–311.

Ehlich, H. (1964). Determinantenabschatzungen für binäre matrizen. *Math. Z.* **83**, 123–132.

El Mossadeq, A. and A. Kobilinsky (1992). Run orders and quantitative factors in asymmetrical designs. *Appl. Stochastic Models Data Anal.* **8**, 259–281.

Farmakis, N. (1992). On constructibility of A-optimal weighing designs $(n, k, 5)$ when $n \equiv 3(\text{mod } 4)$ and $k = n - 1, n$. *J. Statist. Plann. Inference* **33**, 275–283.

Filliben, J. J. and K. C. Li (1997). A systematic approach to the analysis of complex interaction patterns in two level factorial designs. *Technometrics* **39**, 286–297.

Finney, D. J. (1982). Some enumerations for the 6×6 Latin squares. *Utilitias Math.* **21**, 137–153.

Franklin, M. F. (1984). Constructing tables of minimum aberration p^{n-m} designs. *Technometrics* **26**, 225–232.

Franklin, M. F. (1985). Selecting defining contrasts and confounded effects in p^{n-m} factorial experiments. *Technometrics* **27**, 165–172.

Fries, A. and W. G. Hunter (1980). Minimum aberration 2^{k-p} designs. *Technometrics* **22**, 601–608.

Fujii, Y. (1976). An upper bound of resolution in asymmetrical fractional factorial designs. *Ann. Statist.* **4**, 662–667.

Fujii, Y., T. Namikawa, and S. Yamamoto (1987). On three-symbol orthogonal arrays. Contributed Papers, 46th Session of the International Statistical Institute, 131–132.

Galil, Z. and J. Kiefer (1980). D-optimum weighing designs. *Ann. Statist.* **8**, 1293–1306.

Galil, Z. and J. Kiefer (1982). Construction methods for D-optimum weighing designs when $n \equiv 3(\text{mod } 4)$. *Ann. Statist.* **10**, 502–510.

Ghosh, S. (1996). Sequential assembly of fractions in factorial experiments. In *Handbook of Statistics*, vol. 13 (S. Ghosh and C. R. Rao, eds.), pp. 407–435, Amsterdam: North-Holland.

Ghosh, S. and X. D. Zhang (1987). Two new series of search designs for 3^m factorial experiments. *Utilitas Math.* **32**, 245–254.

Goswami, K. K. and S. Pal (1992). On the construction of orthogonal factorial designs of resolution IV. *Commun. Statist. Theory Meth.* **21**, 3561–3570.

Gupta, B. C. (1990). A survey of search designs for 2^m factorial experiments. In *Probability, Statistics and Design of Experiments* (R. R. Bahadur, ed.), pp. 329–345, New Delhi: Wiley Eastern.

Gupta, S. and K. Chatterjee (1998). Supersaturated designs: A review. *J. Combin. Inf. System Sc.*, to appear.

Gupta, S. and R. Mukerjee (1989). *A Calculus for Factorial Arrangements*. Lecture Notes in Statistics 59. Berlin: Springer-Verlag.

Gupta, V. K. and A. K. Nigam (1985). A class of asymmetrical orthogonal Resolution IV designs. *J. Statist. Plann. Inference* **11**, 381–383.

Gupta, V. K., A. K. Nigam, and A. Dey (1982). Orthogonal main effect plans for asymmetrical factorials. *Technometrics* **24**, 135–137.

Hadley, G. (1962). *Linear Programming*. Reading, MA: Addison Wesley.

Hall, M., Jr. (1986). *Combinatorial Theory*, 2nd ed. New York: Wiley.

Hamada, M. and N. Balakrishnan (1998). Analyzing unreplicated factorial experiments: A review with some new proposals (with discussion). *Statist. Sinica* **8**, 1–41.

Hamada, M. and C. F. J. Wu (1999). *Experiments: Planning, Analysis and Parameter Design Optimization*. New York: Wiley (to appear).

Hedayat, A. S. and H. Pesotan (1990). Strongly threefold orthogonal matrices with statistical applications. *Linear Algebra Appl.* **136**, 1–23.

Hedayat, A. S. and H. Pesotan (1992). Two-level factorial designs for main effects and selected two-factor interactions. *Statist. Sinica* **2**, 453–464.

Hedayat, A. S. and H. Pesotan (1997). Designs for two-level factorial experiments with linear models containing main effects and selected two-factor interactions. *J. Statist. Plann. Inference* **64**, 109–124.

Hedayat, A. S., K. Pu, and J. Stufken (1992). On the construction of asymmetrical orthogonal arrays. *Ann. Statist.* **20**, 2142–2152.

Hedayat, A. S., B. L. Raktoe, and W. T. Federer (1974). On a measure of aliasing due to fitting an incomplete model. *Ann. Statist.* **2**, 650–660.

Hedayat, A. S., E. Seiden, and J. Stufken (1997). On the maximal number of factors and the enumeration of 3-symbol orthogonal arrays of strength 3 and index 2. *J. Statist. Plann. Inference* **58**, 43–63.

Hedayat, A. S. and J. Stufken (1988). Two-symbol orthogonal arrays. In *Optimal Design and Analysis of Experiments* (Y. Dodge et al., eds.), pp. 47–58, Amsterdam: North-Holland.

Hedayat, A. S. and J. Stufken (1989). On the maximum number of constraints in orthogonal arrays. *Ann. Statist.* **17**, 448–451.

Hedayat, A. S., J. Stufken, and G. Su (1996). On difference schemes and orthogonal arrays of strength t. *J. Statist. Plann. Inference* **56**, 307–324.

Hedayat, S., J. Stufken, and G. Su (1997). On the construction and existence of orthogonal arrays with three levels and indexes 1 and 2. *Ann. Statist.* **25**, 2044–2053.

Hedayat, A. S. and W. D. Wallis (1978). Hadamard matrices and their applications. *Ann. Statist.* **6**, 1184–1238.

Hill, H. H. (1960). Experimental designs to adjust for time trends. *Technometrics* **2**, 67–82.

Hinkelmann, K. and O. Kempthorne (1994). *Design and Analysis of Experiments*, vol. 1. New York: Wiley.

Hong, Y. (1986). On the nonexistence of nontrivial perfect e-codes and tight $2e$-designs in Hamming schemes $H(n, q)$ with $e \geq 3$ and $q \geq 3$. *Graphs Combin.* **2**, 145–164.

Hyodo, Y. and M. Kuwada (1994). Analysis of variance of balanced fractional s^m factorial designs of resolution $V_{p,q}$. *J. Statist. Plann. Inference* **38**, 263–277.

Jacroux, M. (1992). A note on the determination and construction of minimal orthogonal main effect plans. *Technometrics* **34**, 92–96.

Jacroux, M. (1993). On the construction of minimal partially replicated orthogonal main effect plans. *Technometrics* **35**, 32–36.

Jacroux, M. (1994). On the construction of trend resistant mixed level factorial run orders. *Ann. Statist.* **22**, 904–916.

Jacroux, M. (1996). On the construction of trend resistant asymmetrical orthogonal arrays. *Statist. Sinica* **6**, 289–297.

Jacroux, M. and R. Saharay (1990). On the construction of trend free row-column 2-level factorial experiments. *Metrika* **37**, 163–180.

Jacroux, M. and R. Saharay (1991). Run orders of trend resistant 2-level factorial designs. *Sankhyā* **B53**, 202–212.

Jacroux, M., C. S. Wong, and J. C. Masaro (1983). On the optimality of chemical balance weighing designs. *J. Statist. Plann. Inference* **8**, 231–240.

John, P. W. M. (1990). Time trends and factorial experiments. *Technometrics* **32**, 275–282.

John, P. W. M., M. E. Johnson, L. M. Moore, and D. Ylvisaker (1995). Minimax distance designs in two-level factorial experiments. *J. Statist. Plann. Inference* **44**, 249–263.

Joiner, B. L. and C. Campbell (1976). Designing experiments when run order is important. *Technometrics* **18**, 249–259.

Jungnickel, D. (1979). On difference matrices, resolvable transversal designs and generalized Hadamard matrices. *Math. Z.* **167**, 49–60.

Kiefer, J. C. (1975). Construction and optimality of generalized Youden designs. In *A Survey of Statistical Design and Linear Models* (J. N. Srivastava, ed.), pp. 333–353, Amsterdam: North-Holland.

Kobilinsky, A. (1985). Confounding in relation to duality of finite Abelian groups. *Linear Algebra Appl.* **70**, 321–347.

Kobilinsky, A. and H. Monod (1995). Juxtaposition of regular factorial designs and the complex linear model. *Scand. J. Statist.* **22**, 223–254.

Kolyva-Machera, F. (1989a). D-optimality in 3^k designs for $N \equiv 1 \bmod 9$ observations. *J. Statist. Plann. Inference* **22**, 95–103.

Kolyva-Machera, F. (1989b). Fractional factorial designs and G-optimality. In *Proc. Fourth Prague Symp. on Asymptotic Statistics*, pp. 349–358, Prague: Charles University Press.

Koukouvinos, C. (1996). Linear models and D-optimal designs for $n \equiv 2 \bmod 4$. *Statist. Probab. Lett.* **26**, 329–332.

Kounias, S. (1977). Optimal 2^k designs of odd and even resolution. In *Recent Developments in Statistics* (Proc. European Meeting, Grenoble, 1976), pp. 501–506, Amsterdam: North-Holland.

Kounias, S. and T. Chadjipantelis (1983). Some D-optimal weighing designs for $n \equiv 3 (\mathrm{mod}\ 4)$. *J. Statist. Plann. Inference* **8**, 117–127.

Kounias, S., M. Lefkopoulou, and C. Bagiatis (1983). G-optimal N observation first order 2^k designs. *Discrete Math.* **46**, 21–31.

Kounias, S. and C. I. Petros (1975). Orthogonal arrays of strength three and four with index unity. *Sankhyā* **B37**, 228–240.

Krafft, O. (1990). Some matrix representations occurring in linear two-factor models. In *Probability, Statistics and Design of Experiments* (R. R. Bahadur, ed.), pp. 461–470, New Delhi: Wiley Eastern.

Krafft, O. and M. Schaefer (1991). E-optimality of a class of saturated main effect plans. *Statistics* **22**, 9–15.

Krehbiel, T. C. and D. A. Anderson (1991). Optimal fractional factorial designs for estimating interactions of one factor with all others. *Commun. Statist. Theory Meth.* **20**, 1055–1072.

Kunert, J. (1997). On the use of the factor sparsity assumption to get an estimate of the variance in saturated designs. *Technometrics* **39**, 81–90.

Kurkjian, B. and M. Zelen (1962). A calculus for factorial arrangements. *Ann. Math. Statist.* **33**, 600–619.

Kurkjian, B. and M. Zelen (1963). Applications of the calculus for factorial arrangements, I. Block and direct product designs. *Biometrika* **50**, 63–73.

Kuwada, M. (1982). On some optimal fractional 2^m factorial designs of resolution V. *J. Statist. Plann. Inference* **7**, 39–48.

Laycock, P. J. and P. J. Rowley (1995). A method for generating and labelling all regular fractions or blocks for q^{n-m} designs. *J. Roy. Statist. Soc.* **B57**, 191–204.

Lewis, S. M. and J. A. John (1976). Testing main effects in fractions of asymmetrical factorial experiments. *Biometrika* **63**, 678–680.

Li, W. W. and C. F. J. Wu (1997). Columnwise–pairwise algorithms with applications to the construction of supersaturated designs. *Technometrics* **39**, 171–179.

Liao, C. T., H. K. Iyer, and D. F. Vecchia (1996). Construction of orthogonal two-level designs of user-specified resolution where $N \neq 2^k$. *Technometrics* **38**, 342–353.

Lin, D. K. J. (1993). A new class of supersaturated designs. *Technometrics* **35**, 28–31.

Lin, D. K. J. (1995). Generating systematic supersaturated designs. *Technometrics* **37**, 213–225.

Lin, D. K. J. and N. R. Draper (1992). Projection properties of Plackett and Burman designs. *Technometrics* **34**, 423–428.

Lin, M. and A. M. Dean (1991). Trend-free block designs for varietal and factorial experiments. *Ann. Statist.* **19**, 1582–1596.

MacWilliams, F. J. and N. J. A. Sloane (1977). *The Theory of Error Correcting Codes*. Amsterdam: North-Holland.

Mandeli, J. P. (1995). Construction of asymmetrical orthogonal arrays having factors with a large non-prime power number of levels. *J. Statist. Plann. Inference* **47**, 377–391.

Margolin, B. H. (1969). Resolution IV fractional factorial designs. *J. Roy. Statist. Soc.* **B31**, 514–523.

Marshall, A. and I. Olkin (1979). *Inequalities: Theory of Majorization and Its Applications*. New York: Academic Press.

Masaro, J. and C. S. Wong (1992). Type I optimal weighing designs when $N \equiv 3 \bmod 4$. *Utilitas Math.* **41**, 97–107.

Masuyama, M. (1957). On difference sets for constructing orthogonal arrays of index two and of strength two. *Rep. Statist. Appl. Res., JUSE* **5**, 27–34.

Meyer, R. D., D. M. Steinberg, and G. Box (1996). Follow-up designs to resolve confounding in multifactor experiments (with discussion). *Technometrics* **38**, 303–332.

Mitchell, T. J. (1974). Computer construction of "D-optimal" first order designs. *Technometrics* **16**, 211–220.

Monette, G. (1983). A measure of aliasing and applications to two-level fractional factorials. *Canadian J. Statist.* **11**, 199–206.

Mood, A. M. (1946). On Hotelling's weighing problem. *Ann. Math. Statist.* **17**, 432–446.

Morgan, J. P. and N. Uddin (1996). Optimal blocked main effects plans with nested rows and columns. *Ann. Statist.* **24**, 1185–1208.

Moyssiadis, C., S. Chadjiconstantinidis, and S. Kounias (1995). A-optimization of exact first order saturated designs for $N \equiv 1$ mod 4 observations. *Linear Algebra Appl.* **216**, 159–176.

Mukerjee, R. (1979). On a theorem by Bush on orthogonal arrays. *Calcutta Statist. Assoc. Bull.* **28**, 169–170.

Mukerjee, R. (1980). Orthogonal fractional factorial plans. *Calcutta Statist. Assoc. Bull.* **29**, 143–160.

Mukerjee, R. (1982). Universal optimality of fractional factorial plans derivable through orthogonal arrays. *Calcutta Statist. Assoc. Bull.* **31**, 63–68.

Mukerjee, R. (1995). On E-optimal fractions of symmetric and asymmetric factorials. *Statist. Sinica* **5**, 515–533.

Mukerjee, R. (1999). On the optimality of orthogonal array plus one run plans. *Ann. Statist.*, to appear.

Mukerjee, R. and K. Chatterjee (1985). Estimability and efficiency in proportional frequency plans. *J. Indian Soc. Agric. Statist.* **37**, 79–87.

Mukerjee, R. and K. Chatterjee (1994). A search procedure using polychotomies for search linear models with positive error variance. *Statistics & Decisions* **12**, 91–103.

Mukerjee, R., K. Chatterjee, and M. Sen (1986). D-optimality of a class of saturated main effect plans and allied results. *Statistics* **17**, 349–355.

Mukerjee, R. and S. Kageyama (1994). On existence of two-symbol complete orthogonal arrays. *J. Comb. Theory* **A66**, 176–181.

Mukerjee, R. and B. K. Sinha (1990). Almost saturated D-optimal main effect plans and allied results. *Metrika* **37**, 301–307.

Mukerjee, R. and C. F. J. Wu (1995). On the existence of saturated and nearly saturated asymmetrical orthogonal arrays. *Ann. Statist.* **23**, 2102–2115.

Mukerjee, R. and C. F. J. Wu (1997a). Minimum aberration designs for mixed factorials in terms of complementary sets. Preprint.

Mukerjee, R. and C. F. J. Wu (1997b). Blocking in regular fractional factorials: A projective geometric approach. Preprint.

Mukhopadhyay, A. C. (1981). Construction of some series of orthogonal arrays. *Sankhyā* **B43**, 81–92.

Nguyen, N. K. (1996a). An algorithmic approach to constructing supersaturated designs. *Technometrics* **38**, 69–73.

Nguyen, N. K. (1996b). A note on the construction of near orthogonal arrays with mixed levels and economic run size. *Technometrics* **38**, 279–283.

Nguyen, N. K. and A. Dey (1989). Computer aided construction of D-optimal 2^m fractional factorial designs of resolution V. *Austral. J. Statist.* **31**, 111–117.

Nguyen, N. K. and A. J. Miller (1997). 2^m fractional factorial designs of resolution V with high A-efficiency, $7 \leq m \leq 10$. *J. Statist. Plann. Inference* **59**, 379–384.

Noda, R. (1979). On orthogonal arrays of strength 4 achieving Rao's bound. *J. London Math. Soc.* **19**, 385–390.

Noda, R. (1986). On orthogonal arrays of strength 3 and 5 achieving Rao's bound. *Graphs Combin.* **2**, 277–282.

Paik, U. B. and W. T. Federer (1973). On construction of fractional replicates and on aliasing schemes. *Ann. Inst. Statist. Math.* **25**, 567–585.

Pesotan, H. and B. L. Raktoe (1985). Some determinant optimality aspects of main effect foldover designs. *J. Statist. Plann. Inference* **11**, 399–410.

Pesotan, H. and B. L. Raktoe (1988). On invariance and randomization in factorial designs with applications to D-optimal main effect designs of the symmetric factorial. *J. Statist. Plann. Inference* **19**, 283–298.

Phillips, J. P. N. (1968). Methods of constructing one-way and factorial designs balanced for trend. *Appl. Statist.* **17**, 162–170.

Plackett, R. L. (1946). Some generalizations in the multifactorial design. *Biometrika* **33**, 328–332.

Plackett, R. L. and J. P. Burman (1946). The design of optimum multifactorial experiments. *Biometrika* **33**, 305–325.

Pukelsheim, F. (1993). *Optimal Design of Experiments*. New York: Wiley.

Raghavarao, D. (1971). *Constructions and Combinatorial Problems in Design of Experiments*. New York: Wiley.

Raghavarao, D., D. A. Anderson, and H. Pesotan (1991). A class of optimally embedded resolution III designs. *Utilitas Math.* **39**, 33–39.

Raktoe, B. L. (1974). A geometrical formulation of an unsolved problem in fractional factorial designs. *Commun. Statist. Theory Meth.* **3**, 959–968.

Raktoe, B. L. (1976). On alias matrices and generalized defining relationships of equi-information factorial experiments. *J. Roy. Statist. Soc.* **B38**, 279–283.

Raktoe, B. L. and W. T. Federer (1970). Characterization of optimal saturated main-effect plans of the 2^n factorial. *Ann. Math. Statist.* **41**, 203–206.

Raktoe, B. L. and W. T. Federer (1973). Balanced optimal saturated main-effect plans of the 2^m factorial and their relation to (v, k, λ) configurations. *Ann. Statist.* **1**, 924–932.

Raktoe, B. L., A. Hedayat, and W. T. Federer (1981). *Factorial Designs*. New York: Wiley.

Rao, C. R. (1947). Factorial experiments derivable from combinatorial arrangements of arrays. *J. Roy. Statist. Soc. suppl.* **9**, 128–139.

Rao, C. R. (1973a). *Linear Statistical Inference and Its Applications* (2nd. ed.). New York: Wiley.

Rao, C. R. (1973b). Some combinatorial problems of arrays and applications to design of experiments. In *A Survey of Combinatorial Theory* (J. N. Srivastava, ed.), pp. 349–359, Amsterdam: North-Holland.

Saha, G. M. (1975). Some results on tactical configurations and related topics. *Utilitas Math.*, **7**, 223–240.

Saha, G. M., B. L. Raktoe, and H. Pesotan (1982). On the problem of augmented fractional factorial designs. *Commun. Statist. Theory Meth.* **11**, 2731–2745.

Sathe, Y. S. and R. G. Shenoy (1989). A-optimal weighing designs when $N \equiv 3 \pmod 4$. *Ann. Statist.* **17**, 1906–1915.

Sathe, Y. S. and R. G. Shenoy (1990). Construction method for some A- and D-optimal weighing designs when $N \equiv 3 \pmod 4$. *J. Statist. Plann. Inference* **24**, 369–375.

Sathe, Y. S. and R. G. Shenoy (1991). Further results on construction methods for some A- and D-optimal weighing designs when $N \equiv 3 \pmod 4$. *J. Statist. Plann. Inference* **28**, 339–352.

Saunders, I. W., J. A. Eccleston, and R. J. Martin (1995). An algorithm for the design of 2^p factorial experiments on continuous processes. *Austral. J. Statist.* **37**, 353–365.

Sawade, K. (1985). A Hadamard matrix of order 268. *Graphs Combin.* **1**, 185–187.

Seberry, J. (1979). Some remarks on generalized Hadamard matrices and theorems of Rajkundlia on SBIBD's. In *Combinatorial Mathematics VI* (A. F. Haradam and W. D. Wallis, eds.), pp. 154–164, Berlin: Springer-Verlag.

Seiden, E. (1954). On the problem of construction of orthogonal arrays. *Ann. Math. Statist.* **25**, 151–156.

Seiden, E. and R. Zemach (1966). On orthogonal arrays. *Ann. Math. Statist.* **37**, 1355–1370.

Shah, B. V. (1958). On balancing in factorial experiments. *Ann. Math. Statist.* **29**, 766–779.

Shah, K. R. and B. K. Sinha (1989). *Theory of Optimal Designs*. Lecture Notes in Statistics 54. Berlin: Springer-Verlag.

Shirakura, T. (1976). Optimal balanced fractional 2^m factorial designs of resolution VII, $6 \leq m \leq 8$. *Ann. Statist.* **4**, 515–531.

Shirakura, T. (1993). Fractional factorial designs of two and three levels. *Discrete Math.* **116**, 99–135.

Shirakura, T., T. Takahashi, and J. N. Srivastava (1996). Searching probabilities for nonzero effects in search designs for the noisy case. *Ann. Statist.* **24**, 2560–2568.

Shirakura, T. and S. Tazawa (1992). A series of search designs for 2^m factorial designs of resolution V which permit search of one or two unknown extra three-factor interactions. *Ann. Inst. Statist. Math.* **44**, 185–196.

Shirakura, T. and W. P. Tong (1996). Weighted A-optimality for fractional 2^m factorial designs of resolution V. *J. Statist. Plann. Inference* **56**, 243–256.

Shrikhande, S. S. (1964). Generalized Hadamard matrices and orthogonal arrays of strength two. *Canadian J. Math.* **16**, 736–740.

Shrikhande, S. S. and Bhagwandas (1969). A note on embedding of orthogonal arrays of strength two. In *Combinatorial Mathematics and Its Applications*, pp. 256–273, Chapel Hill: University of North Carolina Press.

Shrikhande, S. S. and Bhagwandas (1970). A note on embedding of Hadamard matrices. In *Essays in Probability and Statistics* (R. C. Bose et al., eds.), pp. 673–688, Chapel Hill: University of North Carolina Press.

Sinha, B. K. and R. Mukerjee (1982). A note on the universal optimality criterion for full rank models. *J. Statist. Plann. Inference* **7**, 97–100.

Sitter, R. R., J. Chen, and M. Feder (1997). Fractional resolution and minimum aberration in blocked 2^{n-k} designs. *Technometrics* **39**, 382–390.

Sloane, N. J. A. and J. Stufken (1996). A linear programming bound for orthogonal arrays with mixed levels. *J. Statist. Plann. Inference* **56**, 295–305.

Srivastava, J. N. (1970). Optimal balanced 2^m fractional factorial designs. In *Essays in Probability and Statistics* (R. C. Bose et al., eds.), pp. 689–706, Chapel Hill: University of North Carolina Press.

Srivastava, J. N. (1975). Designs for searching non-negligible effects. In *A Survey of Statistical Design and Linear Models* (J. N. Srivastava, ed.), pp. 507–519, Amsterdam: North-Holland.

Srivastava, J. N. (1978). A review of some recent work on discrete optimal factorial designs for statisticians and experimenters. In *Developments in Statistics*, vol. 1 (P. R. Krishnaiah, ed.), pp. 267–329, New York: Academic Press.

Srivastava, J. N. (1987). Advances in the general theory of factorial designs based on partial pencils in Euclidean n-space. *Utilitas Math.* **32**, 75–94.

Srivastava, J. N. (1996). A critique of some aspects of experimental design. In *Handbook of Statistics*, vol. 13 (S. Ghosh and C. R. Rao, eds.), pp. 309–341, Amsterdam: North-Holland.

Srivastava, J. N., D. A. Anderson, and J. Mardekian (1984). Theory of factorial designs of the parallel flats type, I. The coefficient matrix. *J. Statist. Plann. Inference* **9**, 229–252.

Srivastava, J. N. and S. Arora (1991). An infinite series of resolution III. 2 designs for the 2^m factorial experiment. *Discrete Math.* **98**, 35–56.

Srivastava, J. N. and D. V. Chopra (1971a). On the characteristic roots of the information matrix of 2^m balanced factorial designs of resolution V, with applications. *Ann. Math. Statist.* **42**, 722–734.

Srivastava, J. N. and D. V. Chopra (1971b). Balanced optimal 2^m fractional factorial designs of resolution V, $m \leq 6$. *Technometrics* **13**, 257–269.

Srivastava, J. N. and S. Ghosh (1976). A series of 2^m factorial designs of resolution V which allow search and estimation of one extra unknown effect. *Sankhyā* **B38**, 280–289.

Srivastava, J. N. and S. Ghosh (1996). On nonorthogonality and nonoptimality of Addelman's main effect plans satisfying the condition of proportional frequencies. *Statist. Probab. Lett.* **26**, 51–60.

Srivastava, J. N. and J. Li (1996). Orthogonal designs of parallel flats type. *J. Statist. Plann. Inference* **53**, 261–283.

Srivastava, J. N. and D. W. Mallenby (1985). On a decision rule using dichotomies for identifying the nonnegligible parameter in certain linear models. *J. Multivariate Anal.* **16**, 318–334.

Steinberg, D. M. (1988). Factorial experiments with time trends. *Technometrics* **30**, 259–269.

Street, A. P. and D. J. Street (1987). *Combinatorics of Experimental Design*. New York: Claredon Press.

Street, D. J. (1979). Generalized Hadamard matrices, orthogonal arrays and F-squares. *Ars Combin.* **8**, 131–141.

Street, D. J. (1994). Constructions for orthogonal main effect plans. *Utilitas Math.* **45**, 115–123.

Street, D. J. and L. Burgess (1994). A survey of orthogonal main effect plans and related structures. *Congr. Numer.* **99**, 223–239.

Suen, C. (1989a). A class of orthogonal main effect plans. *J. Statist. Plann. Inference* **21**, 391–394.

Suen, C. (1989b). Some resolvable orthogonal arrays with two symbols. *Commun. Statist. Theory Meth.* **18**, 3875–3881.

Suen, C., H. Chen, and C. F. J. Wu (1997). Some identities on q^{n-m} designs with application to minimum aberration designs. *Ann. Statist.* **25**, 1176–1188.

Sun, D. X. and C. F. J. Wu (1994). Interaction graphs for three-level fractional factorial designs. *J. Quality Tech.* **26**, 297–307.

Sun, D. X., C. F. J. Wu, and Y. Chen (1997). Optimal blocking schemes for 2^n and 2^{n-p} designs. *Technometrics* **39**, 298–307.

Tang, B. and C. F. J. Wu (1996). Characterization of minimum aberration 2^{n-k} designs in terms of their complementary designs. *Ann. Statist.* **24**, 2549–2559.

Tang, B. and C. F. J. Wu (1997). A method for constructing supersaturated designs and its $E(s^2)$ optimality. *Canadian J. Statist.* **25**, 191–201.

Vijayan, K. and K. R. Shah (1997). Optimal 12 run designs. Preprint.

Voss, D. T. (1997). Analysis of orthogonal saturated designs. *J. Statist. Plann. Inference*, to appear.

Wang, J. C. (1996). Mixed difference matrices and the construction of orthogonal arrays. *Statist. Probab. Lett.* **28**, 121–126.

Wang, J. C. and C. F. J. Wu (1991). An approach to the construction of asymmetrical orthogonal arrays. *J. Amer. Statist. Assoc.* **86**, 450–456.

Wang, J. C. and C. F. J. Wu (1992). Nearly orthogonal arrays with mixed levels and small runs. *Technometrics* **34**, 409–422.

Wang, J. C. and C. F. J. Wu (1995). A hidden projection property of Plackett-Burman and related designs. *Statist. Sinica* **5**, 235–250.

Wang, P. C. (1990). On the constructions of some orthogonal main effect plans. *Sankhyā* **B52**, 319–323.

Wang, P. C. (1991). Symbol changes and trend-resistance in orthogonal plans of symmetric factorials. *Sankhyā* **B53**, 297–303.

Wang, P. C. (1996). Level changes and trend resistance in $L_N(2^p 4^q)$ orthogonal arrays. *Statist. Sinica* **6**, 471–479.

Webb, S. R. (1968). Non-orthogonal designs of even resolution. *Technometrics* **10**, 291–300.

Wu, C. F. J. (1989). Construction of $2^m 4^n$ designs via a grouping scheme. *Ann. Statist.* **17**, 1880–1885.

Wu, C. F. J. (1993). Construction of supersaturated designs through partially aliased interactions. *Biometrika* **80**, 661–669.

Wu, C. F. J. and Y. Chen (1992). A graph-aided method for planning two-level experiments when certain interactions are important. *Technometrics* **34**, 162–174.

Wu, C. F. J. and R. Zhang (1993). Minimum aberration designs with two-level and four-level factors. *Biometrika* **80**, 203–209.

Wu, C. F. J., R. Zhang, and R. Wang (1992). Construction of asymmetrical orthogonal arrays of the type $OA(s^k, s^m(s^{r_1})^{n_1} \cdots (s^{r_t})^{n_t})$. *Statist. Sinica* **2**, 203–219.

Yamamoto, S., T. Shirakura, and M. Kuwada (1975). Balanced arrays of strength $2l$ and balanced fractional 2^m factorial designs. *Ann. Inst. Statist. Math.* **27**, 143–157.

Yamamoto, S., T. Shirakura, and M. Kuwada (1976). Characteristic polynomials of the information matrices of balanced fractional 2^m factorial designs of higher $(2l + 1)$ resolution. In *Essays in Probability and Statistics* (S. Ikeda et al., eds.), pp. 73–94, Tokyo: Shinko Tsusho.

Yang, C. H. (1968). On designs of maximal $(+1, -1)$ matrices of order $n = 2(\text{mod } 4)$. *Math. Computation* **22**, 174–180.

Zelen, M. (1958). Use of group-divisible designs for confounded asymmetric factorial experiments. *Ann. Math. Statist.* **29**, 22–40.

Index

WILEY SERIES IN PROBABILITY AND STATISTICS
ESTABLISHED BY WALTER A. SHEWHART AND SAMUEL S. WILKS

Editors
Vic Barnett, Noel A. C. Cressie, Nicholas I. Fisher,
Iain M. Johnstone, J. B. Kadane, David G. Kendall, David W. Scott,
Bernard W. Silverman, Adrian F. M. Smith, Jozef L. Teugels;
Ralph A. Bradley, Emeritus, J. Stuart Hunter, Emeritus

Probability and Statistics Section

*ANDERSON · The Statistical Analysis of Time Series
ARNOLD, BALAKRISHNAN, and NAGARAJA · A First Course in Order Statistics
ARNOLD, BALAKRISHNAN, and NAGARAJA · Records
BACCELLI, COHEN, OLSDER, and QUADRAT · Synchronization and Linearity: An Algebra for Discrete Event Systems
BASILEVSKY · Statistical Factor Analysis and Related Methods: Theory and Applications
BERNARDO and SMITH · Bayesian Statistical Concepts and Theory
BILLINGSLEY · Convergence of Probability Measures
BOROVKOV · Asymptotic Methods in Queuing Theory
BOROVKOV · Ergodicity and Stability of Stochastic Processes
BRANDT, FRANKEN, and LISEK · Stationary Stochastic Models
CAINES · Linear Stochastic Systems
CAIROLI and DALANG · Sequential Stochastic Optimization
CONSTANTINE · Combinatorial Theory and Statistical Design
COOK · Regression Graphics
COVER and THOMAS · Elements of Information Theory
CSÖRGŐ and HORVÁTH · Weighted Approximations in Probability Statistics
CSÖRGŐ and HORVÁTH · Limit Theorems in Change Point Analysis
DETTE and STUDDEN · The Theory of Canonical Moments with Applications in Statistics, Probability, and Analysis
DEY and MUKERJEE · Fractional Factorial Plans
*DOOB · Stochastic Processes
DRYDEN and MARDIA · Statistical Analysis of Shape
DUPUIS and ELLIS · A Weak Convergence Approach to the Theory of Large Deviations
ETHIER and KURTZ · Markov Processes: Characterization and Convergence
FELLER · An Introduction to Probability Theory and Its Applications, Volume 1, *Third Edition,* Revised; Volume II, *Second Edition*
FULLER · Introduction to Statistical Time Series, *Second Edition*
FULLER · Measurement Error Models
GHOSH, MUKHOPADHYAY, and SEN · Sequential Estimation
GIFI · Nonlinear Multivariate Analysis
GUTTORP · Statistical Inference for Branching Processes
HALL · Introduction to the Theory of Coverage Processes
HAMPEL · Robust Statistics: The Approach Based on Influence Functions
HANNAN and DEISTLER · The Statistical Theory of Linear Systems
HUBER · Robust Statistics
IMAN and CONOVER · A Modern Approach to Statistics
JUREK and MASON · Operator-Limit Distributions in Probability Theory
KASS and VOS · Geometrical Foundations of Asymptotic Inference

*Now available in a lower priced paperback edition in the Wiley Classics Library.

Probability and Statistics (Continued)

KAUFMAN and ROUSSEEUW · Finding Groups in Data: An Introduction to Cluster Analysis
KELLY · Probability, Statistics, and Optimization
LINDVALL · Lectures on the Coupling Method
McFADDEN · Management of Data in Clinical Trials
MANTON, WOODBURY, and TOLLEY · Statistical Applications Using Fuzzy Sets
MORGENTHALER and TUKEY · Configural Polysampling: A Route to Practical Robustness
MUIRHEAD · Aspects of Multivariate Statistical Theory
OLIVER and SMITH · Influence Diagrams, Belief Nets and Decision Analysis
*PARZEN · Modern Probability Theory and Its Applications
PRESS · Bayesian Statistics: Principles, Models, and Applications
PUKELSHEIM · Optimal Experimental Design
RAO · Asymptotic Theory of Statistical Inference
RAO · Linear Statistical Inference and Its Applications, *Second Edition*
RAO and SHANBHAG · Choquet-Deny Type Functional Equations with Applications to Stochastic Models
ROBERTSON, WRIGHT, and DYKSTRA · Order Restricted Statistical Inference
ROGERS and WILLIAMS · Diffusions, Markov Processes, and Martingales, Volume I: Foundations, *Second Edition;* Volume II: Îto Calculus
RUBINSTEIN and SHAPIRO · Discrete Event Systems: Sensitivity Analysis and Stochastic Optimization by the Score Function Method
RUZSA and SZEKELY · Algebraic Probability Theory
SCHEFFE · The Analysis of Variance
SEBER · Linear Regression Analysis
SEBER · Multivariate Observations
SEBER and WILD · Nonlinear Regression
SERFLING · Approximation Theorems of Mathematical Statistics
SHORACK and WELLNER · Empirical Processes with Applications to Statistics
SMALL and McLEISH · Hilbert Space Methods in Probability and Statistical Inference
STAPLETON · Linear Statistical Models
STAUDTE and SHEATHER · Robust Estimation and Testing
STOYANOV · Counterexamples in Probability
TANAKA · Time Series Analysis: Nonstationary and Noninvertible Distribution Theory
THOMPSON and SEBER · Adaptive Sampling
WELSH · Aspects of Statistical Inference
WHITTAKER · Graphical Models in Applied Multivariate Statistics
YANG · The Construction Theory of Denumerable Markov Processes

Applied Probability and Statistics Section

ABRAHAM and LEDOLTER · Statistical Methods for Forecasting
AGRESTI · Analysis of Ordinal Categorical Data
AGRESTI · Categorical Data Analysis
ANDERSON, AUQUIER, HAUCK, OAKES, VANDAELE, and WEISBERG · Statistical Methods for Comparative Studies
ARMITAGE and DAVID (editors) · Advances in Biometry
*ARTHANARI and DODGE · Mathematical Programming in Statistics
ASMUSSEN · Applied Probability and Queues
*BAILEY · The Elements of Stochastic Processes with Applications to the Natural Sciences
BARNETT and LEWIS · Outliers in Statistical Data, *Third Edition*

*Now available in a lower priced paperback edition in the Wiley Classics Library.

BARTHOLOMEW, FORBES, and McLEAN · Statistical Techniques for Manpower Planning, *Second Edition*
BATES and WATTS · Nonlinear Regression Analysis and Its Applications
BECHHOFER, SANTNER, and GOLDSMAN · Design and Analysis of Experiments for Statistical Selection, Screening, and Multiple Comparisons
BELSLEY · Conditioning Diagnostics: Collinearity and Weak Data in Regression
BELSLEY, KUH, and WELSCH · Regression Diagnostics: Identifying Influential Data and Sources of Collinearity
BHAT · Elements of Applied Stochastic Processes, *Second Edition*
BHATTACHARYA and WAYMIRE · Stochastic Processes with Applications
BIRKES and DODGE · Alternative Methods of Regression
BLOOMFIELD · Fourier Analysis of Time Series: An Introduction
BOLLEN · Structural Equations with Latent Variables
BOULEAU · Numerical Methods for Stochastic Processes
BOX · Bayesian Inference in Statistical Analysis
BOX and DRAPER · Empirical Model-Building and Response Surfaces
BOX and DRAPER · Evolutionary Operation: A Statistical Method for Process Improvement
BUCKLEW · Large Deviation Techniques in Decision, Simulation, and Estimation
BUNKE and BUNKE · Nonlinear Regression, Functional Relations and Robust Methods: Statistical Methods of Model Building
CHATTERJEE and HADI · Sensitivity Analysis in Linear Regression
CHILÈS and DELFINER · Geostatistics: Modeling Spatial Uncertainty
CHOW and LIU · Design and Analysis of Clinical Trials: Concepts and Methodologies
CLARKE and DISNEY · Probability and Random Processes: A First Course with Applications, *Second Edition*
*COCHRAN and COX · Experimental Designs, *Second Edition*
CONOVER · Practical Nonparametric Statistics, *Second Edition*
CORNELL · Experiments with Mixtures, Designs, Models, and the Analysis of Mixture Data, *Second Edition*
*COX · Planning of Experiments
CRESSIE · Statistics for Spatial Data, *Revised Edition*
DANIEL · Applications of Statistics to Industrial Experimentation
DANIEL · Biostatistics: A Foundation for Analysis in the Health Sciences, *Sixth Edition*
DAVID · Order Statistics, *Second Edition*
*DEGROOT, FIENBERG, and KADANE · Statistics and the Law
DODGE · Alternative Methods of Regression
DOWDY and WEARDEN · Statistics for Research, *Second Edition*
DRYDEN and MARDIA · Statistical Shape Analysis
DUNN and CLARK · Applied Statistics: Analysis of Variance and Regression, *Second Edition*
ELANDT-JOHNSON and JOHNSON · Survival Models and Data Analysis
EVANS, PEACOCK, and HASTINGS · Statistical Distributions, *Second Edition*
FLEISS · The Design and Analysis of Clinical Experiments
FLEISS · Statistical Methods for Rates and Proportions, *Second Edition*
FLEMING and HARRINGTON · Counting Processes and Survival Analysis
GALLANT · Nonlinear Statistical Models
GLASSERMAN and YAO · Monotone Structure in Discrete-Event Systems
GNANADESIKAN · Methods for Statistical Data Analysis of Multivariate Observations, *Second Edition*
GOLDSTEIN and LEWIS · Assessment: Problems, Development, and Statistical Issues
GREENWOOD and NIKULIN · A Guide to Chi-Squared Testing
*HAHN · Statistical Models in Engineering

*Now available in a lower priced paperback edition in the Wiley Classics Library.

Applied Probability and Statistics (Continued)

HAHN and MEEKER · Statistical Intervals: A Guide for Practitioners
HAND · Construction and Assessment of Classification Rules
HAND · Discrimination and Classification
HEIBERGER · Computation for the Analysis of Designed Experiments
HINKELMAN and KEMPTHORNE: · Design and Analysis of Experiments, Volume 1: Introduction to Experimental Design
HOAGLIN, MOSTELLER, and TUKEY · Exploratory Approach to Analysis of Variance
HOAGLIN, MOSTELLER, and TUKEY · Exploring Data Tables, Trends and Shapes
HOAGLIN, MOSTELLER, and TUKEY · Understanding Robust and Exploratory Data Analysis
HOCHBERG and TAMHANE · Multiple Comparison Procedures
HOCKING · Methods and Applications of Linear Models: Regression and the Analysis of Variables
HOGG and KLUGMAN · Loss Distributions
HOSMER and LEMESHOW · Applied Logistic Regression
HØYLAND and RAUSAND · System Reliability Theory: Models and Statistical Methods
HUBERTY · Applied Discriminant Analysis
JACKSON · A User's Guide to Principle Components
JOHN · Statistical Methods in Engineering and Quality Assurance
JOHNSON · Multivariate Statistical Simulation
JOHNSON and KOTZ · Distributions in Statistics
Continuous Multivariate Distributions
JOHNSON, KOTZ, and BALAKRISHNAN · Continuous Univariate Distributions, Volume 1, *Second Edition*
JOHNSON, KOTZ, and BALAKRISHNAN · Continuous Univariate Distributions, Volume 2, *Second Edition*
JOHNSON, KOTZ, and BALAKRISHNAN · Discrete Multivariate Distributions
JOHNSON, KOTZ, and KEMP · Univariate Discrete Distributions, *Second Edition*
JUREČKOVÁ and SEN · Robust Statistical Procedures: Aymptotics and Interrelations
KADANE · Bayesian Methods and Ethics in a Clinical Trial Design
KADANE AND SCHUM · A Probabilistic Analysis of the Sacco and Vanzetti Evidence
KALBFLEISCH and PRENTICE · The Statistical Analysis of Failure Time Data
KELLY · Reversability and Stochastic Networks
KHURI, MATHEW, and SINHA · Statistical Tests for Mixed Linear Models
KLUGMAN, PANJER, and WILLMOT · Loss Models: From Data to Decisions
KLUGMAN, PANJER, and WILLMOT · Solutions Manual to Accompany Loss Models: From Data to Decisions
KOVALENKO, KUZNETZOV, and PEGG · Mathematical Theory of Reliability of Time-Dependent Systems with Practical Applications
LAD · Operational Subjective Statistical Methods: A Mathematical, Philosophical, and Historical Introduction
LANGE, RYAN, BILLARD, BRILLINGER, CONQUEST, and GREENHOUSE · Case Studies in Biometry
LAWLESS · Statistical Models and Methods for Lifetime Data
LEE · Statistical Methods for Survival Data Analysis, *Second Edition*
LePAGE and BILLARD · Exploring the Limits of Bootstrap
LINHART and ZUCCHINI · Model Selection
LITTLE and RUBIN · Statistical Analysis with Missing Data
LLOYD · The Statistical Analysis of Categorical Data
MAGNUS and NEUDECKER · Matrix Differential Calculus with Applications in Statistics and Econometrics
MALLER and ZHOU · Survival Analysis with Long Term Survivors
MANN, SCHAFER, and SINGPURWALLA · Methods for Statistical Analysis of Reliability and Life Data

*Now available in a lower priced paperback edition in the Wiley Classics Library.

Applied Probability and Statistics (Continued)

McLACHLAN and KRISHNAN · The EM Algorithm and Extensions
McLACHLAN · Discriminant Analysis and Statistical Pattern Recognition
McNEIL · Epidemiological Research Methods
MEEKER and ESCOBAR · Statistical Methods for Reliability Data
MILLER · Survival Analysis
MONTGOMERY and PECK · Introduction to Linear Regression Analysis, *Second Edition*
MYERS and MONTGOMERY · Response Surface Methodology: Process and Product in Optimization Using Designed Experiments
NELSON · Accelerated Testing, Statistical Models, Test Plans, and Data Analyses
NELSON · Applied Life Data Analysis
OCHI · Applied Probability and Stochastic Processes in Engineering and Physical Sciences
OKABE, BOOTS, and SUGIHARA · Spatial Tesselations: Concepts and Applications of Voronoi Diagrams
PANKRATZ · Forecasting with Dynamic Regression Models
PANKRATZ · Forecasting with Univariate Box-Jenkins Models: Concepts and Cases
PIANTADOSI · Clinical Trials: A Methodologic Perspective
PORT · Theoretical Probability for Applications
PUTERMAN · Markov Decision Processes: Discrete Stochastic Dynamic Programming
RACHEV · Probability Metrics and the Stability of Stochastic Models
RÉNYI · A Diary on Information Theory
RIPLEY · Spatial Statistics
RIPLEY · Stochastic Simulation
ROUSSEEUW and LEROY · Robust Regression and Outlier Detection
RUBIN · Multiple Imputation for Nonresponse in Surveys
RUBINSTEIN · Simulation and the Monte Carlo Method
RUBINSTEIN and MELAMED · Modern Simulation and Modeling
RYAN · Statistical Methods for Quality Improvement
SCHUSS · Theory and Applications of Stochastic Differential Equations
SCOTT · Multivariate Density Estimation: Theory, Practice, and Visualization
*SEARLE · Linear Models
SEARLE · Linear Models for Unbalanced Data
SEARLE, CASELLA, and McCULLOCH · Variance Components
SENNOTT · Stochastic Dynamic Programming and the Control of Queueing Systems
STOYAN, KENDALL, and MECKE · Stochastic Geometry and Its Applications, *Second Edition*
STOYAN and STOYAN · Fractals, Random Shapes and Point Fields: Methods of Geometrical Statistics
THOMPSON · Empirical Model Building
THOMPSON · Sampling
TIJMS · Stochastic Modeling and Analysis: A Computational Approach
TIJMS · Stochastic Models: An Algorithmic Approach
TITTERINGTON, SMITH, and MAKOV · Statistical Analysis of Finite Mixture Distributions
UPTON and FINGLETON · Spatial Data Analysis by Example, Volume 1: Point Pattern and Quantitative Data
UPTON and FINGLETON · Spatial Data Analysis by Example, Volume II: Categorical and Directional Data
VAN RIJCKEVORSEL and DE LEEUW · Component and Correspondence Analysis
WEISBERG · Applied Linear Regression, *Second Edition*
WESTFALL and YOUNG · Resampling-Based Multiple Testing: Examples and Methods for p-Value Adjustment
WHITTLE · Systems in Stochastic Equilibrium
WOODING · Planning Pharmaceutical Clinical Trials: Basic Statistical Principles

*Now available in a lower priced paperback edition in the Wiley Classics Library.

Applied Probability and Statistics (Continued)

WOOLSON · Statistical Methods for the Analysis of Biomedical Data
*ZELLNER · An Introduction to Bayesian Inference in Econometrics

Texts and References Section

AGRESTI · An Introduction to Categorical Data Analysis
ANDERSON · An Introduction to Multivariate Statistical Analysis, *Second Edition*
ANDERSON and LOYNES · The Teaching of Practical Statistics
ARMITAGE and COLTON · Encyclopedia of Biostatistics: Volumes 1 to 6 with Index
BARTOSZYNSKI and NIEWIADOMSKA-BUGAJ · Probability and Statistical Inference
BERRY, CHALONER, and GEWEKE · Bayesian Analysis in Statistics and Econometrics: Essays in Honor of Arnold Zellner
BHATTACHARYA and JOHNSON · Statistical Concepts and Methods
BILLINGSLEY · Probability and Measure, *Second Edition*
BOX · R. A. Fisher, the Life of a Scientist
BOX, HUNTER, and HUNTER · Statistics for Experimenters: An Introduction to Design, Data Analysis, and Model Building
BOX and LUCEÑO · Statistical Control by Monitoring and Feedback Adjustment
BROWN and HOLLANDER · Statistics: A Biomedical Introduction
CHATTERJEE and PRICE · Regression Analysis by Example, *Second Edition*
COOK and WEISBERG · An Introduction to Regression Graphics
COX · A Handbook of Introductory Statistical Methods
DILLON and GOLDSTEIN · Multivariate Analysis: Methods and Applications
DODGE and ROMIG · Sampling Inspection Tables, *Second Edition*
DRAPER and SMITH · Applied Regression Analysis, *Third Edition*
DUDEWICZ and MISHRA · Modern Mathematical Statistics
DUNN · Basic Statistics: A Primer for the Biomedical Sciences, *Second Edition*
FISHER and VAN BELLE · Biostatistics: A Methodology for the Health Sciences
FREEMAN and SMITH · Aspects of Uncertainty: A Tribute to D. V. Lindley
GROSS and HARRIS · Fundamentals of Queueing Theory, *Third Edition*
HALD · A History of Probability and Statistics and their Applications Before 1750
HALD · A History of Mathematical Statistics from 1750 to 1930
HELLER · MACSYMA for Statisticians
HOEL · Introduction to Mathematical Statistics, *Fifth Edition*
HOLLANDER and WOLFE · Nonparametric Statistical Methods, *Second Edition*
HOSMER and LEMESHOW · Applied Survival Analysis: Regression Modeling of Time to Event Data
JOHNSON and BALAKRISHNAN · Advances in the Theory and Practice of Statistics: A Volume in Honor of Samuel Kotz
JOHNSON and KOTZ (editors) · Leading Personalities in Statistical Sciences: From the Seventeenth Century to the Present
JUDGE, GRIFFITHS, HILL, LÜTKEPOHL, and LEE · The Theory and Practice of Econometrics, *Second Edition*
KHURI · Advanced Calculus with Applications in Statistics
KOTZ and JOHNSON (editors) · Encyclopedia of Statistical Sciences: Volumes 1 to 9 wtih Index
KOTZ and JOHNSON (editors) · Encyclopedia of Statistical Sciences: Supplement Volume
KOTZ, REED, and BANKS (editors) · Encyclopedia of Statistical Sciences: Update Volume 1
KOTZ, REED, and BANKS (editors) · Encyclopedia of Statistical Sciences: Update Volume 2
LAMPERTI · Probability: A Survey of the Mathematical Theory, *Second Edition*

*Now available in a lower priced paperback edition in the Wiley Classics Library.

Texts and References (Continued)

LARSON · Introduction to Probability Theory and Statistical Inference, *Third Edition*
LE · Applied Categorical Data Analysis
LE · Applied Survival Analysis
MALLOWS · Design, Data, and Analysis by Some Friends of Cuthbert Daniel
MARDIA · The Art of Statistical Science: A Tribute to G. S. Watson
MASON, GUNST, and HESS · Statistical Design and Analysis of Experiments with Applications to Engineering and Science
MURRAY · X-STAT 2.0 Statistical Experimentation, Design Data Analysis, and Nonlinear Optimization
PURI, VILAPLANA, and WERTZ · New Perspectives in Theoretical and Applied Statistics
RENCHER · Methods of Multivariate Analysis
RENCHER · Multivariate Statistical Inference with Applications
ROSS · Introduction to Probability and Statistics for Engineers and Scientists
ROHATGI · An Introduction to Probability Theory and Mathematical Statistics
RYAN · Modern Regression Methods
SCHOTT · Matrix Analysis for Statistics
SEARLE · Matrix Algebra Useful for Statistics
STYAN · The Collected Papers of T. W. Anderson: 1943–1985
TIERNEY · LISP-STAT: An Object-Oriented Environment for Statistical Computing and Dynamic Graphics
WONNACOTT and WONNACOTT · Econometrics, *Second Edition*

WILEY SERIES IN PROBABILITY AND STATISTICS

ESTABLISHED BY WALTER A. SHEWHART AND SAMUEL S. WILKS

Editors
Robert M. Groves, Graham Kalton, J. N. K. Rao, Norbert Schwarz, Christopher Skinner

Survey Methodology Section

BIEMER, GROVES, LYBERG, MATHIOWETZ, and SUDMAN · Measurement Errors in Surveys
COCHRAN · Sampling Techniques, *Third Edition*
COUPER, BAKER, BETHLEHEM, CLARK, MARTIN, NICHOLLS, and O'REILLY (editors) · Computer Assisted Survey Information Collection
COX, BINDER, CHINNAPPA, CHRISTIANSON, COLLEDGE, and KOTT (editors) · Business Survey Methods
*DEMING · Sample Design in Business Research
DILLMAN · Mail and Telephone Surveys: The Total Design Method
GROVES and COUPER · Nonresponse in Household Interview Surveys
GROVES · Survey Errors and Survey Costs
GROVES, BIEMER, LYBERG, MASSEY, NICHOLLS, and WAKSBERG · Telephone Survey Methodology
*HANSEN, HURWITZ, and MADOW · Sample Survey Methods and Theory, Volume 1: Methods and Applications
*HANSEN, HURWITZ, and MADOW · Sample Survey Methods and Theory, Volume II: Theory

*Now available in a lower priced paperback edition in the Wiley Classics Library.

Survey Methodology (Continued)

KISH · Statistical Design for Research

*KISH · Survey Sampling

LESSLER and KALSBEEK · Nonsampling Error in Surveys

LEVY and LEMESHOW · Sampling of Populations: Methods and Applications, *Third Edition*

LYBERG, BIEMER, COLLINS, de LEEUW, DIPPO, SCHWARZ, TREWIN (editors) · Survey Measurement and Process Quality

SIRKEN, HERRMANN, SCHECHTER, SCHWARZ, TANUR, and TOURANGEAU (editors) · Cognition and Survey Research

SKINNER, HOLT, and SMITH · Analysis of Complex Surveys

*Now available in a lower priced paperback edition in the Wiley Classics Library.